PROGRESS IN BRAIN RESEARCH

VOLUME 179

GENETIC MODELS OF SCHIZOPHRENIA

Other volumes in PROGRESS IN BRAIN RESEARCH

Volume 144: The Roots of Visual Awareness, by C.A. Heywood, A.D. Milner and C. Blakemore (Eds.) – 2004, ISBN 0-444-50978-X.
Volume 145: Acetylcholine in the Cerebral Cortex, by L. Descarries, K. Krnjevic and M. Steriade (Eds.) – 2004, ISBN 0-444-51125-3.
Volume 146: NGF and Related Molecules in Health and Disease, by L. Aloe and L. Calzà (Eds.) – 2004, ISBN 0-444-51472-4.
Volume 147: Development, Dynamics and Pathology of Neuronal Networks: From Molecules to Functional Circuits, by J. Van Pelt, M. Kamermans, C.N. Levelt, A. Van Ooyen, G.J.A. Ramakers and P.R. Roelfsema (Eds.) – 2005, ISBN 0-444-51663-8.
Volume 148: Creating Coordination in the Cerebellum, by C.I. De Zeeuw and F. Cicirata (Eds.) – 2005, ISBN 0-444-51754-5.
Volume 149: Cortical Function: A View from the Thalamus, by V.A. Casagrande, R.W. Guillery and S.M. Sherman (Eds.) – 2005, ISBN 0-444-51679-4.
Volume 150: The Boundaries of Consciousness: Neurobiology and Neuropathology, by Steven Laureys (Ed.) – 2005, ISBN 0-444-51851-7.
Volume 151: Neuroanatomy of the Oculomotor System, by J.A. Büttner-Ennever (Ed.) – 2006, ISBN 0-444-51696-4.
Volume 152: Autonomic Dysfunction after Spinal Cord Injury, by L.C. Weaver and C. Polosa (Eds.) – 2006, ISBN 0-444-51925-4.
Volume 153: Hypothalamic Integration of Energy Metabolism, by A. Kalsbeek, E. Fliers, M.A. Hofman, D.F. Swaab, E.J.W. Van Someren and R.M. Buijs (Eds.) – 2006, ISBN 978-0-444-52261-0.
Volume 154: Visual Perception, Part 1, Fundamentals of Vision: Low and Mid-Level Processes in Perception, by S. Martinez-Conde, S.L. Macknik, L.M. Martinez, J.M. Alonso and P.U. Tse (Eds.) – 2006, ISBN 978-0-444-52966-4.
Volume 155: Visual Perception, Part 2, Fundamentals of Awareness, Multi-Sensory Integration and High-Order Perception, by S. Martinez-Conde, S.L. Macknik, L.M. Martinez, J.M. Alonso and P.U. Tse (Eds.) – 2006, ISBN 978-0-444-51927-6.
Volume 156: Understanding Emotions, by S. Anders, G. Ende, M. Junghofer, J. Kissler and D. Wildgruber (Eds.) – 2006, ISBN 978-0-444-52182-8.
Volume 157: Reprogramming of the Brain, by A.R. Møller (Ed.) – 2006, ISBN 978-0-444-51602-2.
Volume 158: Functional Genomics and Proteomics in the Clinical Neurosciences, by S.E. Hemby and S. Bahn (Eds.) – 2006, ISBN 978-0-444-51853-8.
Volume 159: Event-Related Dynamics of Brain Oscillations, by C. Neuper and W. Klimesch (Eds.) – 2006, ISBN 978-0-444-52183-5.
Volume 160: GABA and the Basal Ganglia: From Molecules to Systems, by J.M. Tepper, E.D. Abercrombie and J.P. Bolam (Eds.) – 2007, ISBN 978-0-444-52184-2.
Volume 161: Neurotrauma: New Insights into Pathology and Treatment, by J.T. Weber and A.I.R. Maas (Eds.) – 2007, ISBN 978-0-444-53017-2.
Volume 162: Neurobiology of Hyperthermia, by H.S. Sharma (Ed.) – 2007, ISBN 978-0-444-51926-9.
Volume 163: The Dentate Gyrus: A Comprehensive Guide to Structure, Function, and Clinical Implications, by H.E. Scharfman (Ed.) – 2007, ISBN 978-0-444-53015-8.
Volume 164: From Action to Cognition, by C. von Hofsten and K. Rosander (Eds.) – 2007, ISBN 978-0-444-53016-5.
Volume 165: Computational Neuroscience: Theoretical Insights into Brain Function, by P. Cisek, T. Drew and J.F. Kalaska (Eds.) – 2007, ISBN 978-0-444-52823-0.
Volume 166: Tinnitus: Pathophysiology and Treatment, by B. Langguth, G. Hajak, T. Kleinjung, A. Cacace and A.R. Møller (Eds.) – 2007, ISBN 978-0-444-53167-4.
Volume 167: Stress Hormones and Post Traumatic Stress Disorder: Basic Studies and Clinical Perspectives, by E.R. de Kloet, M.S. Oitzl and E. Vermetten (Eds.) – 2008, ISBN 978-0-444-53140-7.
Volume 168: Models of Brain and Mind: Physical, Computational and Psychological Approaches, by R. Banerjee and B.K. Chakrabarti (Eds.) – 2008, ISBN 978-0-444-53050-9.
Volume 169: Essence of Memory, by W.S. Sossin, J.-C. Lacaille, V.F. Castellucci and S. Belleville (Eds.) – 2008, ISBN 978-0-444-53164-3.
Volume 170: Advances in Vasopressin and Oxytocin – From Genes to Behaviour to Disease, by I.D. Neumann and R. Landgraf (Eds.) – 2008, ISBN 978-0-444-53201-5.
Volume 171: Using Eye Movements as an Experimental Probe of Brain Function — A Symposium in Honor of Jean Büttner-Ennever, by Christopher Kennard and R. John Leigh (Eds.) – 2008, ISBN 978-0-444-53163-6.
Volume 172: Serotonin–Dopamine Interaction: Experimental Evidence and Therapeutic Relevance, by Giuseppe Di Giovanni, Vincenzo Di Matteo and Ennio Esposito (Eds.) – 2008, ISBN 978-0-444-53235-0.
Volume 173: Glaucoma: An Open Window to Neurodegeneration and Neuroprotection, by Carlo Nucci, Neville N. Osborne, Giacinto Bagetta and Luciano Cerulli (Eds.) – 2008, ISBN 978-0-444-53256-5.
Volume 174: Mind and Motion: The Bidirectional Link Between Thought and Action, by Markus Raab, Joseph G. Johnson and Hauke R. Heekeren (Eds.) – 2009, ISBN 978-0-444-53356-2.
Volume 175: Neurotherapy: Progress in Restorative Neuroscience and Neurology — Proceedings of the 25th International Summer School of Brain Research, held at the Royal Netherlands Academy of Arts and Sciences, Amsterdam, The Netherlands, August 25–28, 2008, by J. Verhaagen, E.M. Hol, I. Huitinga, J. Wijnholds, A.A. Bergen, G.J. Boer, D.F. Swaab (Eds.) – 2009, ISBN 978-0-12-374511-8.
Volume 176: Attention, by Narayanan Srinivasan (Ed.) – 2009, ISBN 978-0-444-53426-2.
Volume 177: Coma Science: Clinical and Ethical Implications, by S. Laureys, N.D. Schiff and A.M. Owen (Eds.)–2009, ISBN 978-0-444-53432-3.
Volume 178: Cultural Neuroscience: Cultural Influences on Brain Function, by J.Y. Chiao (Ed.)–2009, ISBN 978-0-444-53361-6.

PROGRESS IN BRAIN RESEARCH

VOLUME 179

GENETIC MODELS OF SCHIZOPHRENIA

EDITED BY

AKIRA SAWA

Departments of Psychiatry and Neuroscience, Johns Hopkins University School of Medicine, Baltimore, MD, USA

ELSEVIER

AMSTERDAM – BOSTON – HEIDELBERG – LONDON – NEW YORK – OXFORD
ĆPARIS – SAN DIEGO – SAN FRANCISCO – SINGAPORE – SYDNEY – TOKYO

Elsevier
360 Park Avenue South, New York, NY 10010-1710
Linacre House, Jordan Hill, Oxford OX2 8DP, UK
Radarweg 29, PO Box 211, 1000 AE Amsterdam, The Netherlands

First edition 2009

British Library Cataloguing in Publication Data
A catalogue record for this book is available from the British Library

Library of Congress Cataloging-in-Publication Data
A catalog record for this book is available from the Library of Congress

ISBN: 978-0-444-53430-9 (this volume)
ISSN: 0079-6123 (Series)

For information on all Elsevier publications
visit our website at elsevierdirect.com

Printed and bound in Great Britain

09 10 11 12 13 10 9 8 7 6 5 4 3 2 1

List of Contributors

A.F.T. Arnsten, Department of Neurobiology, Yale Medical School, New Haven, CT, USA
N.J. Brandon, Wyeth Research, Department of Neuroscience, NJ, USA
R. DiLeone, Department of Psychiatry, Yale University School of Medicine, New Haven, CT, USA
R. Fukumura, RIKEN BioResource Center, Tsukuba, Ibaraki, Japan
K. Furukubo-Tokunaga, Graduate School of Life and Environmental Sciences, University of Tsukuba, Tennodai, Tsukuba, Japan
Y. Gondo, RIKEN BioResource Center, Tsukuba, Ibaraki, Japan
A. Hayashi-Takagi, Department of Psychiatry and Behavioral Sciences, Johns Hopkins University School of Medicine, Baltimore, MD, USA
H. Jaaro-Peled, Department of Psychiatry and Behavioral Sciences, Johns Hopkins University School of Medicine, Baltimore, MD, USA
A. Kamiya, Department of Psychiatry and Behavioral Sciences, Johns Hopkins University School of Medicine, Baltimore, MD, USA
C. Kellendonk, Department of Psychiatry, Department of Pharmacology, New York, NY, USA
M.P. Kelly, Wyeth Research, Department of Neuroscience, NJ, USA
J.A. Morris, Program in Human Molecular Genetics, Department of Pediatrics, Feinberg School of Medicine, Children's Memorial Research Center, Northwestern University, Chicago, IL, USA
M.V. Pletnikov, Departments of Psychiatry and Neuroscience, Johns Hopkins University School of Medicine, Baltimore, MD, USA
A.J. Ramsey, University of Toronto, Department of Pharmacology and Toxicology, Toronto, ON, Canada
A. Sawa, Departments of Psychiatry and Neuroscience, Johns Hopkins University School of Medicine, Baltimore, MD, USA
A.J. Seshadri, Department of Psychiatry and Behavioral Sciences, Johns Hopkins University School of Medicine, Baltimore, MD, USA
A.A. Simen, Department of Psychiatry, Yale University School of Medicine, New Haven, CT, USA
K. Talbot, Center for Neurobiology and Behavior, Department of Psychiatry, University of Pennsylvania, Philadelphia, PA, USA

Preface

This volume focuses on how genetic models for schizophrenia, that is, manipulation in genetic susceptibility factors for the disease, have potential in opening a new window of better understanding of etiology-relevant mechanisms. In addition, these models can provide a promising novel platform for drug screening toward the treatment of the disease. In particular, this volume introduces novel techniques to generate mouse models for schizophrenia and refers to the use of drosophila, zebrafish, and primates for modeling the disease, in addition to rodents. Representative models for schizophrenia currently available in mice are also systematically introduced, which include mice with modulation of neurotransmission (glutamate and dopamine receptors, respectively) and key intracellular signaling (Disrupted in Schizophrenia, Dysbindin, and phosphodiesterase) possibly associated with neurotransmission, together with genetic support.

Key features

- Leading authors describe the most updated strategies to address genetic manipulation for modeling schizophrenia, including their advantages and limitations.
- Investigators working on various levels of organisms/animals from drosophila, zebra fish, rodents, to primates discuss the potential to generate genetic models for schizophrenia toward mechanistic understanding of the disease as well as translational use, such as drug screening.
- Leading authors of this field introduce representative animal models for schizophrenia at present.

Akira Sawa

Contents

SECTION I

Overview

A. Sawa (Ed.)
Progress in Brain Research, Vol. 179
ISSN 0079-6123

CHAPTER 1

Genetic animal models for schizophrenia: advantages and limitations of genetic manipulation in drosophila, zebrafish, rodents, and primates

Akira Sawa*

Departments of Psychiatry and Neuroscience, Johns Hopkins University School of Medicine, Baltimore, MD, USA

Abstract: Schizophrenia is a debilitating mental illness in which major initial risks of the disease during neurodevelopment may disturb postnatal brain maturation, which results in onset after puberty. Family, twin, and adoption studies have suggested an important role for genetic factors in the etiology of schizophrenia. To address the etiology-associated mechanisms and disease course, use of genetic models, that is, manipulation of genetic susceptibility factors, is currently considered to be a powerful tool for biological studies. In this manuscript, advantages and possible limitations in manipulating genetic susceptibility factors for schizophrenia toward modeling the disease are discussed. In addition to mouse models, the potential to use drosophila, zebrafish, and primates is underscored.

Keywords: schizophrenia; gene; animal model

Introduction

Schizophrenia is a debilitating mental illness in which genetic factors are known to play a role. None of the factors causes the disease by itself, but a combination of genetic factors together with environmental impact results in manifestation of the symptoms (Sawa and Snyder, 2002). For the past 10 years, promising genetic susceptibility factors for schizophrenia have been found (Harrison and Weinberger, 2005). Whole genome association studies and work on copy number variations are further discovering new susceptibility factors (Cichon et al., 2009; Williams et al., 2009). In physical disorders and other brain disorders studied in the area of neurology, genetic factors have been utilized to build animal models for diseases (Wong et al., 2002). In analogy to these successful preceding works, many investigators are currently trying to develop genetic models for schizophrenia. In this chapter, the advantages and limitations of genetic models for schizophrenia will be overviewed.

Advantage of genetic animal models

Many epidemiological studies have suggested that initial brain insults associated with schizophrenia occur mainly during early neurodevelopment. Nonetheless, the real onset of the disease is in young adulthood. These observations indicate that postnatal brain maturation may play an

*Corresponding author.
Tel.: +1 410 955 4726; Fax: +1 410 614 1792;
E-mail: asawa1@jhmi.edu

DOI: 10.1016/S0079-6123(09)17901-3

important role in the pathology of schizophrenia (Jaaro-Peled et al., 2009).

Administration of compounds that can induce psychosis in humans, such as phencyclidine, into adult animals has been widely used for modeling schizophrenia (Mouri et al., 2007). Although these models are useful in recapitulating the pathophysiology of schizophrenia in part, it is very difficult to mimic the long-term pathological changes leading from disease-associated insults (triggers, risk factors) during pre- and peri-natal stages to the onset of the disease in young adulthood. To overcome this limitation, introduction of brain lesions by toxins during early development has been considered as an alternative strategy for modeling schizophrenia (Lipska et al., 1993; Talamini et al., 1998). A major concern in this second strategy is that these lesions may not reflect the disease etiology. Therefore, genetic models for schizophrenia, including genetically engineered mice in which genetic susceptibility factor(s) for the disease is (are) modulated, are currently underscored to be a promising novel methodology.

Use of genetic model animals has at least two significant advantages. First, as described above, these will contribute to understanding of the pathology of the disease, including the course and mechanisms relevant to disease etiologies. Indeed, major genetic susceptibility factors for schizophrenia play key roles during early development, which suggests that these factors may be tightly associated with the cause of the disease. Second, such models can be utilized for translational means, such as compound screening for future therapeutic strategy toward schizophrenia. Therefore, high throughput readouts that reflect human pathology and phenotypes may be expected in animal models. Are behavioral assays, however, the most appropriate readouts to evaluate these models? In genetic animal models for Alzheimer's disease, senile plaques or β-amyloid plaques are used as a major readout to evaluate the models. Considering this successful precedence, it may be important to consider objective biomarkers in the study of animal models for schizophrenia. One promising approach may be to utilize decrease in immunoreactivity of parvalbumin, a marker for the fast spiking interneurons, which is known in both autopsied brains from patients with schizophrenia and some genetic animal models for schizophrenia (Hikida et al., 2007; Lisman et al., 2008; Shen et al., 2008). It may be important to cultivate more varieties of molecular biomarkers for schizophrenia. For this purpose, use of human-derived neurons, such as olfactory neurons obtained via nasal biopsy and induced pluripotent stem cell-originated neurons, may be useful resources in the future (Sawa and Cascella, 2009; Takahashi et al., 2007).

Possible limitations of genetic animal models

As described above, schizophrenia is not caused by any single factor; instead, a combination of several genetic and environmental factors results in the disease. A major limitation in most technologies for genetic manipulation in animals is that only one gene can be modified per generation, except by utilizing laborious cross-breeding experiments, whereas modulation of more than one gene may be expected to more faithfully mimic the etio-pathological condition of schizophrenia. How can we overcome this dilemma? Mouse models generated by *in utero* gene transfer or by stereotaxic injection of virus-mediated expression or RNAi constructs, through which we can manipulate expression of more than one gene simultaneously, may open a new window toward overcoming this limitation (Kamiya et al., 2008). Although these mice are not genetic models in a strict sense (because the manipulation is not heritable), these strategies are very important to model schizophrenia by utilizing information about genetic susceptibility factors.

The other difficulty in using genetic information in modeling schizophrenia is that we do not know how disease-associated genetic variations play roles in the pathology. It is unlikely that complete loss-of-function and simple gain-of-function are elicited by such genetic variations. Instead, impairment of specific isoform(s) of the susceptibility factors or partial loss-of-function is frequently suggested to be a mechanism (Tan et al., 2007). In this sense, conventional knockout or transgenic mice may have some limitations in modeling the disease. Techniques that introduce mutations causing minor functional impairments in the target genes or modulations of gene expression in a context-dependent manner may be useful to address this issue. Therefore, models with genetic mutations induced by chemical mutagens and conditional/inducible genetic models are considered to be promising (Gondo, 2008; Li et al., 2007; Pletnikov et al., 2008).

Potential in using drosophila, zebrafish, and primates as models for schizophrenia

Two major goals in building model organisms/animals for schizophrenia are to address molecular mechanisms and to utilize them for translational means, such as drug screening. In order to address neuronal connectivity, deficits in which may underlie the pathology of the disease, drosophila and zebrafish are frequently used in basic neuroscience. Therefore, efforts in modulating schizophrenia-associated genes in these organisms have been started in several laboratories (Drerup et al., 2009; Sawamura et al., 2008; Wood et al., 2009). In addition, these organisms are potentially useful in high throughput screening, as far as straightforward readouts relevant to the disease are available.

Schizophrenia may be able to be modeled in lower organisms and rodents when we focus mainly on translatable (particularly neuropathological, molecular, and physiological) markers. Nonetheless, the question of how to model high brain functions, probably more specific to primates, still remains unanswered. To address this question, use of primates needs to be considered, with discussion on technical and ethical viewpoints. A recent report on a genetically engineered marmoset, a type of non-human primate, may be a breakthrough toward this goal.

Concluding remarks

In this short review, advantages and limitations of genetic models for schizophrenia, or manipulation in genetic susceptibility factors for schizophrenia, are discussed. Many types of animals and organisms, such as rodents (mice and rats), drosophila, zebrafish, and primates, can potentially be utilized toward the goal.

Acknowledgments

I thank Saurav Seshadri for critical reading of this manuscript. I appreciate Yukiko Lema for organizing the figures and manuscript. This work is supported by US Public Heath Service Grant MH-069853, Silvio O. Conte Center grant MH-084018, MH-088753, and foundation grants from Stanley, RUSK, as well as NARSAD.

References

Cichon, S., Craddock, N., Daly, M., Faraone, S. V., Gejman, P. V., Kelsoe, J., et al. (2009). Genomewide association studies: History, rationale, and prospects for psychiatric disorders. *American Journal of Psychiatry*, *166*, 540–556.

Drerup, C. M., Wiora, H. M., Topczewski, J., & Morris, J. A. (2009). Disc1 regulates foxd3 and sox10 expression, affecting neural crest migration and differentiation. *Development*, *136*, 2623–2632.

Gondo, Y. (2008). Trends in large-scale mouse mutagenesis: From genetics to functional genomics. *Nature Reviews Genetics*, *9*, 803–810.

Harrison, P. J., & Weinberger, D. R. (2005). Schizophrenia genes, gene expression, and neuropathology: On the matter of their convergence. *Molecular Psychiatry*, *10*, 40–68.

Hikida, T., Jaaro-Peled, H., Seshadri, S., Oishi, K., Hookway, C., Kong, S., et al. (2007). Dominant-negative DISC1 transgenic mice display schizophrenia-associated phenotypes detected by measures translatable to humans. *Proceedings of the National Academy of Sciences United States of America*, *104*, 14501–14506.

Jaaro-Peled, H., Hayashi-Takagi, A., Seshadri, S., Kamiya, A., Brandon, N. J., & Sawa, A. (2009). Neurodevelopmental mechanisms of schizophrenia: Understanding disturbed postnatal brain maturation through neuregulin-1-ErbB4 and DISC1. *Trends in Neurosciences*, *32*, 485–495.

Kamiya, A., Tan, P. L., Kubo, K., Engelhard, C., Ishizuka, K., Kubo, A., et al. (2008). Recruitment of PCM1 to the centrosome by the cooperative action of DISC1 and BBS4: A candidate for psychiatric illnesses. *Archives of General Psychiatry*, *65*, 996–1006.

Li, W., Zhou, Y., Jentsch, J. D., Brown, R. A., Tian, X., Ehninger, D., et al. (2007). Specific developmental disruption of disrupted-in-schizophrenia-1 function results in schizophrenia-related phenotypes in mice. *Proceedings of the National Academy of Sciences United States of America, 104*, 18280–18285.

Lipska, B. K., Jaskiw, G. E., & Weinberger, D. R. (1993). Postpubertal emergence of hyperresponsiveness to stress and to amphetamine after neonatal excitotoxic hippocampal damage: A potential animal model of schizophrenia. *Neuropsychopharmacology, 9*, 67–75.

Lisman, J. E., Coyle, J. T., Green, R. W., Javitt, D. C., Benes, F. M., Heckers, S., et al. (2008). Circuit-based framework for understanding neurotransmitter and risk gene interactions in schizophrenia. *Trends in Neurosciences, 31*, 234–242.

Mouri, A., Noda, Y., Enomoto, T., & Nabeshima, T. (2007). Phencyclidine animal models of schizophrenia: Approaches from abnormality of glutamatergic neurotransmission and neurodevelopment. *Neurochemistry International, 51*, 173–184.

Pletnikov, M. V., Ayhan, Y., Nikolskaia, O., Xu, Y., Ovanesov, M. V., Huang, H., et al. (2008). Inducible expression of mutant human DISC1 in mice is associated with brain and behavioral abnormalities reminiscent of schizophrenia. *Molecular Psychiatry, 13*, 173–186.

Sawa, A., & Cascella, N. G. (2009). Peripheral olfactory system for clinical and basic psychiatry: A promising entry point to the mystery of brain mechanism and biomarker identification in schizophrenia. *American Journal of Psychiatry, 166*, 137–139.

Sawa, A., & Snyder, S. H. (2002). Schizophrenia: Diverse approaches to a complex disease. *Science, 296*, 692–695.

Sawamura, N., Ando, T., Maruyama, Y., Fujimuro, M., Mochizuki, H., Honjo, K., et al. (2008). Nuclear DISC1 regulates CRE-mediated gene transcription and sleep homeostasis in the fruit fly. *Molecular Psychiatry, 13*, 1138–1148.

Shen, S., Lang, B., Nakamoto, C., Zhang, F., Pu, J., Kuan, S.L., et al. (2008). Schizophrenia-related neural and behavioral phenotypes in transgenic mice expressing truncated Disc1. *Journal of Neuroscience, 28*, 10893–10904.

Takahashi, K., Tanabe, K., Ohnuki, M., Narita, M., Ichisaka, T., Tomoda, K., et al. (2007). Induction of pluripotent stem cells from adult human fibroblasts by defined factors. *Cell, 131*, 861–872.

Talamini, L. M., Koch, T., Ter Horst, G. J., & Korf, J. (1998). Methylazoxymethanol acetate-induced abnormalities in the entorhinal cortex of the rat; parallels with morphological findings in schizophrenia. *Brain Research, 789*, 293–306.

Tan, W., Wang, Y., Gold, B., Chen, J., Dean, M., Harrison, P.J., et al. (2007). Molecular cloning of a brain-specific, developmentally regulated neuregulin 1 (NRG1) isoform and identification of a functional promoter variant associated with schizophrenia. *Journal of Biological Chemistry, 282*, 24343–24351.

Williams, H. J., Owen, M. J., & O'Donovan, M. C. (2009). Schizophrenia genetics: New insights from new approaches. *British Medical Bulletin, 91*, 61–74.

Wong, P. C., Cai, H., Borchelt, D. R., & Price, D. L. (2002). Genetically engineered mouse models of neurodegenerative diseases. *Nature Neuroscience, 5*, 633–639.

Wood, J. D., Bonath, F., Kumar, S., Ross, C. A., & Cunliffe, V.T. (2009). Disrupted-in-schizophrenia 1 and neuregulin 1 are required for the specification of oligodendrocytes and neurones in the zebrafish brain. *Human Molecular Genetics, 18*, 391–404.

SECTION II

Novel Techniques to Generate Genetic Models

A. Sawa (Ed.)
Progress in Brain Research, Vol. 179
ISSN 0079-6123

CHAPTER 2

Animal models for schizophrenia via *in utero* gene transfer: understanding roles for genetic susceptibility factors in brain development

Atsushi Kamiya*

Department of Psychiatry and Behavioral Sciences, Johns Hopkins University School of Medicine, Baltimore, MD, USA

Abstract: Genetic disturbances of brain development may underlie the pathophysiology of schizophrenia. Recent advances in molecular neurobiology suggest that some genetic risk factors for schizophrenia have multiple roles in various brain regions depending on the developmental stage. Furthermore, these factors are likely to act synergistically or epistatically in common molecular pathways, possibly contributing to disease pathology. Thus, a technique that can manipulate the expression of more than one gene simultaneously in animal models is necessary to address such molecular pathways. To produce such animal models, *in utero* gene transfer technique is one useful method. Given that plasmid-based cell-type-specific and inducible gene expression systems are now available, combining these technologies and *in utero* gene transfer opens a new window to examine the functional role of genetic risk factors for schizophrenia by conducting multiple-gene targeting in a spatial and temporal manner. The utility of animal models produced by *in utero* gene transfer will also be expected to be evaluated in terms of functional and behavioral outcomes after puberty, which may be associated with schizophrenia pathology.

Keywords: *in utero* gene transfer; brain development; animal model; genetic factor; schizophrenia

Introduction

Disturbances in neuronal circuit formation during brain development are believed to underlie the pathology of schizophrenia (Lewis and Levitt, 2002; Rapoport et al., 2005). Consistent with this notion, accumulating evidence from neuropathological examinations and brain imaging studies suggest a variety of anatomical and functional disturbances in many brain regions, including the cerebral cortex, hippocampus, thalamus, and amygdala in patients with schizophrenia. This concept is further supported by the fact that many genetic risk factors for schizophrenia, such as Disrupted-in-Schizophrenia-1 (DISC1) and Neuregulin-1, play key roles in neurodevelopment (Harrison and Weinberger, 2005; Jaaro-Peled et al., 2009; Owen et al., 2005). Thus, it is important to examine how disturbances of such risk factors impact neuronal circuit formation during brain development. Some of them are likely to function not only in early developmental processes (i.e. pre- and perinatal stages), but also in

*Corresponding author.
Tel.: +1 41 0502 0060; Fax: +1 41 0614 1792;
E-mail: akamiya1@jhmi.edu

DOI: 10.1016/S0079-6123(09)17902-5

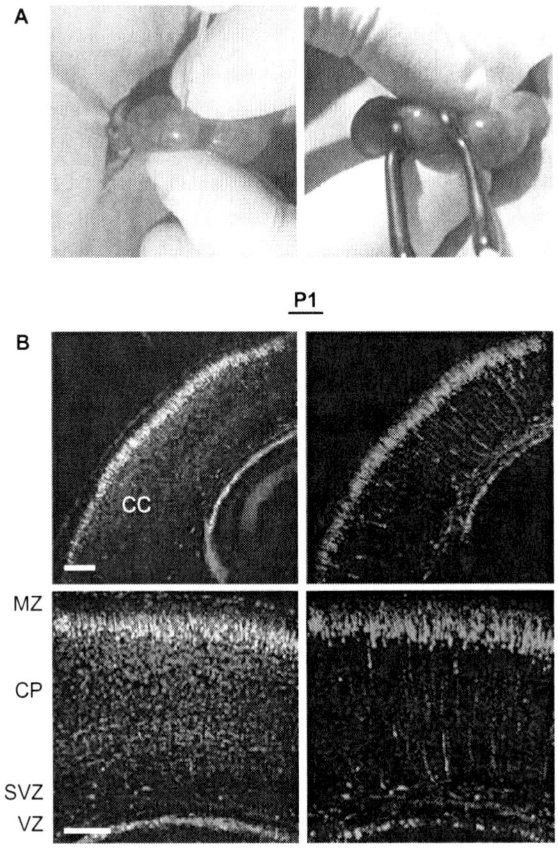

Fig. 1. In utero gene transfer for gene targeting in developing brain. (A) Schematic representation of injection of plasmid constructs into lateral ventricle following by their delivery into ventricular zone via electroporation with forceps-type electrodes. (B) Representative coronal images of brains with GFP expression at postnatal day 1 (P1) after introduction of GFP expression construct at embryonic day 15 (E15). Most GFP-positive cells are in the upper side of cortical plate (CP) which forms layer II/III of cerebral cortex (upper panels, low magnification; bottom panels, high magnification). CC, cerebral cortex; MZ, marginal zone; SVZ, subventricular zone; VZ, ventricular zone. Red, Nucleus. Scale bars, 100 μm.

late developmental stages (i.e. childhood and adolescence) and even in adulthood when their roles may or may not be the same as those in developmental stages. Interestingly, these factors likely act synergistically or epistatically in common molecular pathways during brain development, perhaps leading to susceptibility for schizophrenia by shared pathological mechanisms (Harrison and Weinberger, 2005; Jaaro-Peled et al., 2009; Owen et al., 2005). Therefore, a technique that can manipulate expression of more than one gene simultaneously in a spatial and temporal manner is necessary to address such molecular pathways, which may contribute to the disease pathology.

In this respect, the *in utero* gene transfer technique is a promising methodology. This technique allows for the expression of more than one target gene to be modulated by introduction of expression and/or short hairpin RNA (shRNA) constructs in the developing brain. In this technique, plasmid expression vectors are injected into the lateral ventricles of the embryonic brain through the uterine wall and are introduced into the ventricular zone by electroporation (Tabata and Nakajima, 2001) (Fig. 1). Since the electroporated embryos develop normally *in utero*, we can characterize them at any developmental stages, and even at the adult stage. There are

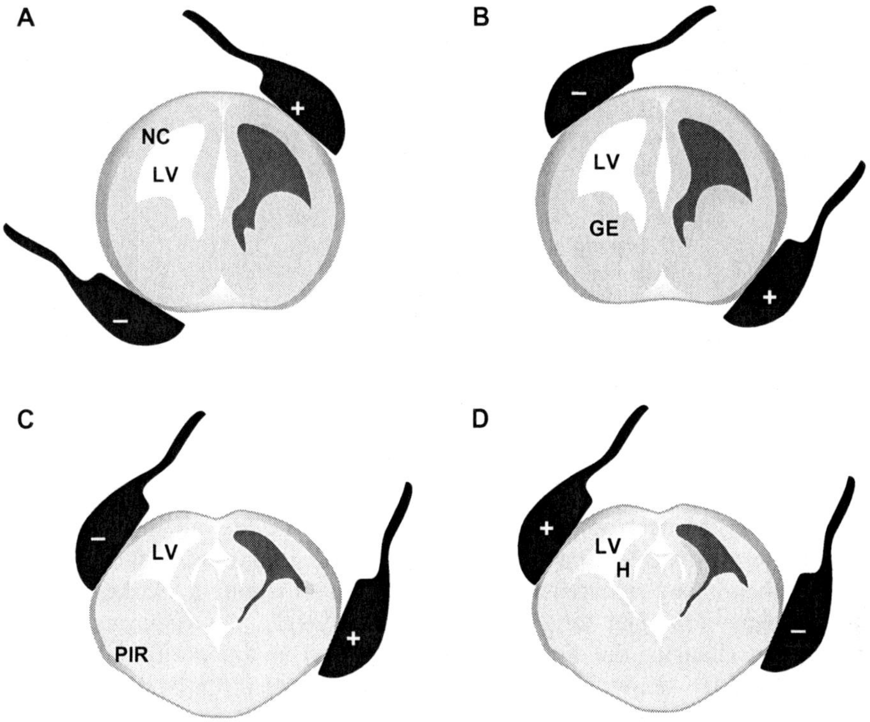

Fig. 2. Region-specific gene targeting by in utero gene transfer. Schematic representation of in utero gene transfer by changing the direction of electroporation for gene targeting into specific regions. (A) Neocortical neuroepithelium is targeted by dorsal lateral placement of the positive (+) electrode. (B) Ganglionic eminence is targeted by ventral lateral placement of the positive electrode at about 30° outward angle from the horizontal plane. (C) Lateroventral pallial neuroepithelium is targeted by ventral lateral placement of the positive electrode at about 15° outward angle from the horizontal plane. (D) Ammonic neuroepithelium is targeted by dorsal lateral placement of the positive electrode on side without plasmids injection. NC, neocortex; GE, ganglionic eminence; PIR, piriform cortex; H, hippocampus.

several advantages in using this method, in comparison with conventional genetically engineered animal models. First, gene targeting can be conducted in specific cell types in restricted brain regions without causing embryonic lethality that may result from the lack of a crucial molecule during development. Second, we can save the time required to produce animal models, because the method is much simpler and quicker. Furthermore, the specific functional interaction between a target molecule and its binding partner(s) can be examined through co-electroporation of multiple constructs. Specifically, rescue experiments may provide crucial information regarding the impact of specific allelic variants that increase risk to disease. In this chapter, we overview the current progress of this technology and discuss its feasibility for addressing the role of genetic risk factors in molecular pathways underlying the pathology of schizophrenia in the context of brain development.

Region-specific targeting by *in utero* gene transfer

The proper development of prefrontal cortex, together with the subcortical regions of the limbic system, including the hippocampus and amygdala, is important for mediating higher brain functioning, such as cognition, memory, and emotion. Each of these regions is implicated in the pathophysiology of schizophrenia (Harrison and Weinberger, 2005). Although many types of genetically engineered animal models including inducible and conditional systems have been established, region-specific gene targeting has not

Table 1. Cell-type-specific promoters

Promoters	Targeted cell types	References
BLBP	Radial glia cells	Gal et al. (2006)
GLAST	Radial glia cells	Gal et al. (2006)
Tα1	Neuronal progenitor cells, immature neurons	Gal et al. (2006)
Nestin	Neuronal progenitor cells, neural stem cells	Miyagi et al. (2004)
Sox2	Neuronal progenitor cells, neural stem cells	Miyagi et al. (2004)
DCX	Postmitotic, immature neurons	Wang et al. (2007)
NeuroD	Postmitotic, immature neurons	Yokota et al. (2007)
CDK5	Postmitotic, immature neurons	Wang et al. (2007)

Many cell-type-specific promoters are now available for gene targeting into specific cell types. By delivering expression constructs with cell-specific promoters, targeted gene-mediated function can be addressed in a specific cell population. BLBP, brain lipid binding protein; GLAST, glutamate transporter gene; Tα1, tubulin α-1; DCX, doublecortin; NeuroD, neurogenic differentiation 1; CDK5, cyclin-dependent kinase 5.

yet been well developed. In utero gene transfer technique was originally applied for modulating target gene expression in the developing parietal cortex of mouse embryos (Fukuchi-Shimogori and Grove, 2001). Since then, many researchers have developed this technique for gene targeting in various brain regions by changing the direction of electroporation (Fig. 2). By simply altering the position of the electrodes, target gene expression can be manipulated in the progenitor cells of specific regions of the brain, including the neocortical neuroepithelium, ammonic neuroepithelium, and lateroventral pallial neuroepithelium at embryonic stages, from which the cerebral cortex, hippocampus, as well as piriform cortex and amygdala are developed, respectively (Bai et al., 2008; Nakahira and Yuasa, 2005; Remedios et al., 2007; Tabata and Nakajima, 2001). Of note, progenitor cells in the ganglionic eminence can also be spatially manipulated, making it possible to analyze specific functions of target genes in interneurons in the cerebral cortex (Borrell et al., 2005).

Cell-type-specific targeting by *in utero* gene transfer

Many types of cells are reportedly impaired at functional and morphological levels in the brains of patients with schizophrenia. Smaller size and/or disarray of pyramidal neurons in hippocampus and prefrontal cortex have been repeatedly reported (Benes et al., 1991; Pierri et al., 2001). Impaired dendritic arborization and orientation as well as reduced dendritic spin density were found using Golgi staining (Glantz and Lewis, 2000). In addition to the abnormalities in glutamatergic pyramidal neurons, reduction of expression level of parvalbumin, a marker for the fast spiking GABAergic interneurons, was reproducibly reported in the prefrontal cortex (Lewis et al., 2004). More recently, oligodendrocyte abnormalities were also demonstrated by brain imaging, microarray, and morphometric analysis (Uranova et al., 2007). Thus, investigation on the functional implication of genetic risk factors in specific cell types is very important. There are two approaches to address this issue. First, changing the timing of electroporation is a simple and useful methodology. For example, electroporation directed into progenitor cells in the neocortical ventricular zone at embryonic day 12 (E12) allows us to modulate gene expression in the cells which typically terminate in layer V/VI and differentiate into pyramidal neurons, whereas cells manipulated by electroporation in the same region at E15 terminate at layer II/III, based on inside-out pattern of cortical layer formation (Langevin et al., 2007). When gene targeting is conducted in the same region at E18, astrocytes are mainly targeted, consistent with the notion that astrocytes originating from the subventricular zone are produced at late embryonic stages and early postnatal stages in rat [E17 to postnatal day 14 (P14)] (Sauvageot and Stiles, 2002). Furthermore, since some genetic risk factors have multiple roles at different developmental stages, timing of gene targeting is crucial for segregating their function in specific cellular events. One example is the case of DISC1, which plays roles in cell proliferation, neuronal migration, and dendritic development (Kamiya et al., 2005; Mao et al., 2009). We have previously reported that knockdown of DISC1 expression by RNAi at E14.5 elicits an impairment of radial neuronal migration and dendritic arborization in the developing cerebral cortex (Kamiya et al., 2005), whereas Mao et al. (2009) recently found that knockdown of DISC1 expression at E13 leads to cell proliferation deficit, clearly indicating the importance of the timing of gene targeting for the study of multifunctional molecules. Although we cannot entirely segregate these functional aspects of DISC1 by changing the timing, a second approach, that is cell-type-specific gene expression driven by diverse promoters, may aid in addressing these questions (Table 1). Given that expression constructs with these promoters become activated in specific cell types, such as radial glia cells, neuronal progenitor cells, and postmitotic immature neurons, cell-type-specific functional role of genetic risk factors can be addressed. To our knowledge, *in utero* gene transfer with single shRNA vector system combined with such cell-type-specific promoters is not yet established. Nonetheless, combination of Cre/LoxP recombination-mediated inducible microRNAs system with cell-type-specific promoters can be useful for loss-of-function experiments in specific cell types (Matsuda and Cepko, 2007). This will be discussed further below.

Multiple-gene targeting: feasibility for studying disease pathways

Schizophrenia is likely a polygenetic disorder with multiple genetic risk factors likely acting synergistically or epistatically (Harrison and Weinberger, 2005; Jaaro-Peled et al., 2009). Since *in utero* gene transfer technique can modulate the expression of more than one gene simultaneously, this technology is particularly useful for studying molecular disease pathways in which several risk factors may converge. In fact, we have previously reported a possible convergent molecular pathway at the centrosome in the developing cerebral cortex where DISC1, PCM1 (another risk factor for schizophrenia), and BBS4, a causative gene for Bardet–Biedl syndrome (BBS) that frequently accompanies cognitive deficits and psychosis, seem to functionally interact with one another (Kamiya et al., 2008). We found that suppression of PCM1 leads to neuronal migration defects, which are phenocopied by the suppression of either DISC1 or BBS4, and are exacerbated by the concomitant suppression of both DISC1 and BBS4. The phenotype induced by RNAi can be further evaluated by rescue experiments. Shu et al. (2004) reported that overexpression of lissencephaly 1 protein (LIS1) partially rescues the impairment of migration caused by loss of function of nuclear distribution element-like (NDEL1) which is known as a major interactor for LIS1, but not that of dynein, whereas overexpression of LIS1 does not rescue the phenotypes by NDEL1 RNAi, suggesting that NDEL1 and LIS1 cooperatively function in the dynein-mediated common pathway for neuronal positioning. By utilizing the same strategy, Young-Pearse et al. (2007) reported functional interaction of β-amyloid precursor protein (APP) with Disabled-1 (Dab1) for neuronal migration by showing that overexpression of Dab1 partially rescued the migration defects caused by APP knockdown. Rescue experiments can also be useful for functional interpretation of protein binding sites and crucial sites for cellular signaling cascades regulating brain development, such as phosphorylation sites. For example, our unpublished data suggest that protein interaction of DISC1 with NDEL1 is crucial for neuronal migration, because overexpression of a DISC1 short fragment (amino acids 788-849), which include amino acids 802-835 of DISC1, that was previously reported as minimal binding domain to NDEL1 (Kamiya et al., 2006), impaired neuronal positioning in developing cerebral cortex.

Conditional targeting system

One major limitation of *in utero* gene transfer technique was that modulation of gene expression may not be precisely temporally controlled. As we discussed above, many genetic factors for schizophrenia have roles at different timings during neurodevelopment (Harrison and Weinberger, 2005; Jaaro-Peled et al., 2009), such that DISC1 has roles in cell proliferation, neuronal migration, and dendritic development/spine formation in the developing cerebral cortex as well as synapse function in cerebral cortex and positioning of newborn neurons in the dentate gyrus at the adult stage (Duan et al., 2007; Jaaro-Peled et al., 2009; Kamiya et al., 2005; Mao et al., 2009). Therefore, it is critical to segregate each molecule's functions in the developmental stages as well as the adult stage in order to address these mechanisms precisely. To overcome this limitation, an inducible gene expression system has been recently applied to *in utero* gene transfer (Matsuda and Cepko, 2007). In this system, a plasmid construct in which Cre recombinase is expressed in response to intraperitoneally injected 4-hydroxytamoxifen (4-OHT) and a target gene or microRNAs expression construct where a stop codon flanked by two loxP sites downstream from the promoter region is contained, are co-electroporated into embryonic brains. Expression of target gene or microRNAs is induced by tamoxifen-mediated Cre recombination mechanism. By utilizing this system, LoTurco and colleagues reported that conditional re-expression of doublecortin (DCX) at P0 normalized impaired neuronal migration elicited by the conventional knockdown of DCX at E14 (Manent et al., 2009). It is noteworthy that re-expression of DCX ameliorated the susceptibility to convulsant-induced seizures in DCX knockdown animal at P30, supporting the utility of *in utero* gene transfer not only for studying on the functional role of target gene in molecular and anatomical levels in brain development, but also for examining how genetic deficits in developmental stages may affect brain functions.

Future directions

Our ultimate goal is to understand the molecular pathways underlying schizophrenia pathology that link between roles of genetic risk factors in brain development and functional abnormalities in neuronal circuits as well as the resultant behavioral phenotypes. Although *in utero gene* transfer technique is a promising methodology for the analysis of polygenic component in vivo, it is unclear whether animals manipulated by this technique could be used for functional and behavioral examination after developmental stage, in particular after puberty when onset of schizophrenia is frequently observed in humans. Nonetheless, DCX knockdown using *in utero* electroporation represents susceptibility to a drug-induced epileptic seizures which frequently appears in patients with DCX mutations (Manent et al., 2009), suggesting that outcomes of *in utero* manipulation are likely to be able to be detected after puberty in functional and behavioral assessments in schizophrenia studies. Finally, gene manipulation by *in utero* gene transfer technique also can be applied in a number of genetically engineered mice models which are already available. In particular, introduction of Cre recombinase expression construct under the control of cell-type-specific promoters by *in utero* electroporation into conditional genetic animal models will be an innovative and attractive approach to segregate a specific functional domain of genetic risk factors responsible for disease-related phenotypes in a spatial and temporal manner.

Acknowledgments

We thank Dr. Tracy L. Young-Pearse for critical reading and Ms. Yukiko Lema for preparation of the figures. We thank Drs. Kenichiro Kubo and Kazunori Nakajima who have provided data presented here. This work was supported by grants from NARSAD and S-R Foundation.

References

Bai, J., Ramos, R. L., Paramasivam, M., Siddiqi, F., Ackman, J.B., & LoTurco, J. J. (2008). The role of DCX and LIS1 in migration through the lateral cortical stream of developing forebrain. *Developmental Neuroscience, 30*, 144–156.

Benes, F. M., Sorensen, I., & Bird, E. D. (1991). Reduced neuronal size in posterior hippocampus of schizophrenic patients. *Schizophrenia Bulletin, 17*, 597–608.

Borrell, V., Yoshimura, Y., & Callaway, E. M. (2005). Targeted gene delivery to telencephalic inhibitory neurons by directional *in utero* electroporation. *Journal of Neuroscience Methods, 143*, 151–158.

Duan, X., Chang, J. H., Ge, S., Faulkner, R. L., Kim, J. Y., Kitabatake, Y., et al. (2007). Disrupted-in-Schizophrenia 1 regulates integration of newly generated neurons in the adult brain. *Cell, 130*, 1146–1158.

Fukuchi-Shimogori, T., & Grove, E. A. (2001). Neocortex patterning by the secreted signaling molecule FGF8. *Science, 294*, 1071–1074.

Gal, J. S., Morozov, Y. M., Ayoub, A. E., Chatterjee, M., Rakic, P., & Haydar, T. F. (2006). Molecular and morphological heterogeneity of neural precursors in the mouse neocortical proliferative zones. *Journal of Neuroscience, 26*, 1045–1056.

Glantz, L. A., & Lewis, D. A. (2000). Decreased dendritic spine density on prefrontal cortical pyramidal neurons in schizophrenia. *Archives of General Psychiatry, 57*, 65–73.

Harrison, P. J., & Weinberger, D. R. (2005). Schizophrenia genes, gene expression, and neuropathology: On the matter of their convergence. *Molecular Psychiatry, 10*, 40–68.

Jaaro-Peled, H., Hayashi-Takagi, H., Seshadri, S., Kamiya, A., Brandon, J.N., & Sawa, A. (2009). Neurodevelopmental mechanisms of schizophrenia: Understanding disturbed postnatal brain maturation through NRG1–ERBB4 and DISC1. Trends in Neurosciences, *32*(9), 485–495.

Kamiya, A., Kubo, K., Tomoda, T., Takaki, M., Youn, R., Ozeki, Y., Sawamura, N., et al. (2005). A schizophrenia-associated mutation of DISC1 perturbs cerebral cortex development. *Nature Cell Biology, 7*, 1167–1178.

Kamiya, A., Tomoda, T., Chang, J., Takaki, M., Zhan, C., Morita, M., et al. (2006). DISC1-NDEL1/NUDEL protein interaction, an essential component for neurite outgrowth, is modulated by genetic variations of DISC1. *Human Molecular Genetics, 15*, 3313–3323.

Kamiya, A., Tan, P. L., Kubo, K., Engelhard, C., Ishizuka, K., Kubo, A., et al. (2008). Recruitment of PCM1 to the centrosome by the cooperative action of DISC1 and BBS4: A candidate for psychiatric illnesses. *Archives in General Psychiatry, 65*, 996–1006.

Langevin, L. M., Mattar, P., Scardigli, R., Roussigne, M., Logan, C., Blader, P., et al. (2007). Validating *in utero* electroporation for the rapid analysis of gene regulatory elements in the murine telencephalon. *Developmental Dynamics, 236*, 1273–1286.

Lewis, D. A., & Levitt, P. (2002). Schizophrenia as a disorder of neurodevelopment. *Annual Review of Neuroscience, 25*, 409–432.

Lewis, D. A., Volk, D. W., & Hashimoto, T. (2004). Selective alterations in prefrontal cortical GABA neurotransmission in schizophrenia: A novel target for the treatment of working memory dysfunction. *Psychopharmacology (Berl), 174*, 143–150.

Manent, J. B., Wang, Y., Chang, Y., Paramasivam, M., & LoTurco, J. J. (2009). Dcx reexpression reduces subcortical band heterotopia and seizure threshold in an animal model of neuronal migration disorder. *Nature Medicine, 15*, 84–90.

Mao, Y., Ge, X., Frank, C. L., Madison, J. M., Koehler, A. N., Doud, M. K., et al. (2009). Disrupted in schizophrenia 1 regulates neuronal progenitor proliferation via modulation of GSK3beta/beta-catenin signaling. *Cell, 136*, 1017–1031.

Matsuda, T., & Cepko, C. L. (2007). Controlled expression of transgenes introduced by in vivo electroporation. *Proceedings of the National Academy of Sciences United States of America, 104*, 1027–1032.

Miyagi, S., Saito, T., Mizutani, K., Masuyama, N., Gotoh, Y., Iwama, A., et al. (2004). The Sox-2 regulatory regions display their activities in two distinct types of multipotent stem cells. *Molecular Cell Biology, 24*, 4207–4220.

Nakahira, E., & Yuasa, S. (2005). Neuronal generation, migration, and differentiation in the mouse hippocampal primoridium as revealed by enhanced green fluorescent protein gene transfer by means of *in utero* electroporation. *Journal of Comparative Neurology, 483*, 329–340.

Owen, M. J., Craddock, N., & O'Donovan, M. C. (2005). Schizophrenia: Genes at last? *Trends in Genetics, 21*, 518–525.

Pierri, J. N., Volk, C. L., Auh, S., Sampson, A., & Lewis, D. A. (2001). Decreased somal size of deep layer 3 pyramidal neurons in the prefrontal cortex of subjects with schizophrenia. *Archives in General Psychiatry, 58*, 466–473.

Rapoport, J. L., Addington, A. M., Frangou, S., & Psych, M. R. (2005). The neurodevelopmental model of schizophrenia: Update 2005. *Molecular Psychiatry, 10*, 434–449.

Remedios, R., Huilgol, D., Saha, B., Hari, P., Bhatnagar, L., Kowalczyk, T., Hevner, R. F., et al. (2007). A stream of cells migrating from the caudal telencephalon reveals a link between the amygdala and neocortex. *Nature Neuroscience, 10*, 1141–1150.

Sauvageot, C. M., & Stiles, C. D. (2002). Molecular mechanisms controlling cortical gliogenesis. *Current Opinion in Neurobiology, 12*, 244–249.

Shu, T., Ayala, R., Nguyen, M. D., Xie, Z., Gleeson, J. G., & Tsai, L. H. (2004). Ndel1 operates in a common pathway with LIS1 and cytoplasmic dynein to regulate cortical neuronal positioning. *Neuron, 44*, 263–277.

Tabata, H., & Nakajima, K. (2001). Efficient *in utero* gene transfer system to the developing mouse brain using electroporation: Visualization of neuronal migration in the developing cortex. *Neuroscience, 103*, 865–872.

Uranova, N. A., Vostrikov, V. M., Vikhreva, O. V., Zimina, I.S., Kolomeets, N. S., & Orlovskaya, D. D. (2007). The role of oligodendrocyte pathology in schizophrenia. *International Journal of Neuropsychopharmacology, 10*, 537–545.

Wang, X., Qiu, R., Tsark, W., & Lu, Q. (2007). Rapid promoter analysis in developing mouse brain and genetic labeling of young neurons by doublecortin-DsRed-express. *Journal of Neuroscience Research, 85*, 3567–3573.

Yokota, Y., Ring, C., Cheung, R., Pevny, L., & Anton, E. S. (2007). Nap1-regulated neuronal cytoskeletal dynamics is essential for the final differentiation of neurons in cerebral cortex. *Neuron, 54*, 429–445.

Young-Pearse, T. L., Bai, J., Chang, R., Zheng, J. B., LoTurco, J. J., & Selkoe, D. J. (2007). A critical function for beta-amyloid precursor protein in neuronal migration revealed by *in utero* RNA interference. *Journal of Neuroscience, 27*, 14459–14469.

A. Sawa (Ed.)
Progress in Brain Research, Vol. 179
ISSN 0079-6123

CHAPTER 3

Gene manipulation with stereotaxic viral infection for psychiatric research: Spatiotemporal components for schizophrenia

Anupamaa J. Seshadri and Akiko Hayashi-Takagi*

Department of Psychiatry and Behavioral Sciences, Johns Hopkins University School of Medicine, Baltimore, MD, USA

Abstract: Schizophrenia (SZ) is a highly polygenic disease with strong genetic predisposition. Although genetic susceptibility factors for SZ are likely to have an influence in some brain regions and related neural circuits during neurodevelopment, direct proof for spatiotemporal causality in the development of SZ, and the alteration of what gene function at what brain region during what developmental stage, remains to be elucidated. Gene manipulation by viral vector stereotaxically injected into a specific brain region is now becoming available for psychiatric research. This technique has several advantages, *e.g.*, the exceptional spatiotemporal control, simultaneous manipulation of multiple genes, and its simple protocol. These properties can make this technique one of the most valuable approaches for research in SZ, which is a complex brain disorder with multifactorial, genetic, and developmental features. This review summarizes the benefits and actual use of this technique together with discussion of spatiotemporal aspect for SZ.

Keywords: schizophrenia (SZ); a polygenic disease; stereotaxic viral injection; spatiotemporal gene manipulation; susceptible neural circuits for SZ

Introduction

Schizophrenia (SZ) is a chronic and devastating mental illness affecting over 50 million people worldwide (Gottesman, 1991; Lewis and Lieberman, 2000). Despite its frequency and severity, its pathogenesis is still poorly understood mainly because of methodological limitations in handling the human brain (Sawa and Snyder, 2002). In order to get better insight into the living brain, numerous animal models have been generated (Kellendonk et al., 2009). Traditionally, pharmacological animal models have been developed based on the manipulation of neurotransmitter systems believed to be involved in SZ (Davis et al., 1991; Geyer, 1998; Goff and Coyle, 2001; Moghaddam, 2003). Perhaps the best known pharmacological model involves administration of amphetamine and phencyclidine to manipulate dopaminergic and glutamatergic transmissions (Javitt and Zukin, 1991; Pierce and Kalivas, 1997). In parallel, other animal models based on the concept of neurodevelopmental insults as risk factors for SZ have been made in

*Corresponding author.
Tel.: +1 410 614 1780; Fax: +410 614 1792;
E-mail: ahayash4@jhmi.edu

DOI: 10.1016/S0079-6123(09)17903-7

order to obtain prominent aspects of SZ that are not addressed by pharmacological models (Weinberger, 1987). The neonatal lesion model is one of the established lesion models for SZ, in which the production of neonatal lesions results in increased dopaminergic hyperresponsivity to psychostimulants only after puberty (Lipska et al., 1993).

The most important advantage in using animal models, however, may be that genetic–phenotypic association can be directly addressed by generating genetically engineered animals (Kellendonk et al., 2009). This advantage holds considerable promise for SZ research, because human genetic research strongly indicates genetic influences on development of SZ (McClellan et al., 2007; Owen et al., 2005; Stefansson et al., 2009). Genetically engineered mice have been widely used in addressing the role of genes in a psychiatric context. Nonetheless, embryonic lethality is often reported as a result of gene targeting. In other cases, in contrast, no apparent or un-expected phenotype is observed, which is sometimes considered to be caused by redundancy or compensatory mechanisms by other genes. To overcome these drawbacks, conditional knockout mice with use of the Cre-*loxP* system and the tetracycline responsive element have been utilized, which allow for spatiotemporally controlled genetic modifications (Tsien et al., 1996). These are, no doubt, outstanding techniques which have provided numerous findings, but to establish such conditional mice can be costly, and usually takes a couple of years.

Alternative techniques for gene manipulation with spatiotemporal control are *in utero* gene transfer (see the other section in this volume) and stereotaxic viral injection. Whereas *in utero* gene transfer can only be used in an embryo during the last half of pregnancy (Tabata and Nakajima, 2001), stereotaxic viral injection can handle a broad range of postnatal animals (Cetin et al., 2006). While each method has advantages and limitations (Table 1), SZ is now considered a disorder of neurodevelopment in a certain set of neural circuits (Arnold et al., 2005; Cannon et al., 2003; Lewis and Levitt, 2002; McGlashan and Hoffman, 2000; Rapoport et al., 2005; Winterer and Weinberger, 2004); thus stereotaxic injection appears to be one of the most valuable approaches for SZ research. Thus, in this review, we focus on the application of stereotaxic virus injection related to psychiatric research, together with a discussion of what neural circuit might be altered, which in turn might correspond to the SZ manifestation.

Why is this technique useful for psychiatric research?

It is now widely accepted that SZ is a polygenic disease, and that the combination of multifactorial components, especially genetic components, is required for SZ development (Harrison and Weinberger, 2005). Although the individual who later develops SZ has the genetic predisposition congenitally, the symptoms are not obvious until their onset during adolescence (Lewis and Lieberman, 2000). The plausible explanation for this paradox, a genetic predisposition with a late adolescence onset, is that (1) SZ-related genes are operating more strongly in late adolescence to complete the neural circuit, and/or (2) there is an additive effect of environmental insult during adolescence on the pre-existing predisposition of genetic vulnerability. While this synergistic combination of genetic predisposition and environmental insult is assumed to be the cause of SZ, direct examination and understanding of what gene alteration occurs at what brain region during which developmental stage as a cause for SZ remains to be elucidated. The biggest advantage of the stereotaxic injection technique may be the exceptional spatiotemporal control (Cetin et al., 2006), which can be useful to answer these questions. Before describing the actual application of this technique for SZ research, we briefly summarize the findings showing what spatiotemporal components are considered to be relevant to the development of SZ.

Temporal aspects of SZ etiology and/or pathophysiology

Births during winter/spring and in urban areas have been reported to be more frequent among

Table 1. Comparison among currently available animal models for psychiatric research

	Reflect genetic component	Multiple gene manipulation	Cell-type-specific control	Spatial control	Temporal control	Developmental stage addressed	Required time for a protocol
Pharmacological model	–	NA	–	–	++	Postnatal	Hours to weeks
Lesion model	–	NA	–	++	++	Postnatal	Hours
Conventional gene engineering mouse	+	+	–	–	–	From fertilization	1 year
Conditional gene engineering mouse	+	+	++	++	++	From fertilization	Years
In utero gene transfer	+	++	±	+	++	Last half of prenatal	Hours
Stereotaxic viral injection	+	++	++	++	++	Postnatal	Hours
	– : no +: yes	NA: not applicable +: a couple genes ++: several genes	– : no ±: poor +: possible ++ : established				

individuals with SZ than in the general population (Mortensen et al., 1999; Torrey et al., 1997). The explanation may be that deleterious factors, including infection, toxins, or malnutrition, are more common in urban populations and that in winter, they may affect brain development. Indeed, numbers of studies reported that virus infections such as influenza are risk factors for SZ (Brown et al., 2004; Buka et al., 2008; Carter, 2008; Mednick et al., 1988; Yolken and Torrey, 1995). SZ has also been associated with perinatal processes such as obstetrical complications (Cannon et al., 2002). Brain damage from hypoxia is, at least in part, considered as a main mechanism underlying this effect.

Likewise, various insults during postnatal development are also reported to increase the risk of SZ. Urban upbringing is indicated as a possible risk factor (Lewis et al., 1992; Pedersen and Mortensen, 2001). In fact, moving into a more urban area or into a more rural area increases or decreases risk for SZ, respectively. The rationale for this might be the increased chance of childhood illness or head trauma. Social causes and increased incidence of stressful life events in urban life could also be involved. Indeed, cannabis use and stressful life events before adolescence have been correlated with an increase in the risk of development of SZ (Andreasson et al., 1987; Weber et al., 2008). These suggest that a variety of pre-, peri-, and postnatal factors confer risk for SZ, probably by cross-talking with genes related to neurodevelopment, and this eventually alters neural circuits during the neurodevelopmental process.

Spatial aspects of SZ etiology and/or pathophysiology

The precise brain regions and related neural circuits involved in SZ remain to be elucidated. In recent years, various lines of study have detected structural and functional changes in some brain areas in SZ. Postmortem studies, although many of them have conflicting results, have suggested that there is no gliosis and no cell death, but there are differing accounts of reduced grey matter volume in such regions as the right basal ganglia, right prefrontal cortex, right insular cortex, right temporal cortex, thalamus, and bilateral cingulate cortex (Harrison, 1999). Decreases in the density of dendritic spines in the prefrontal cortex, temporal cortex, and hippocampus, but not in visual cortex, were also reported, suggesting that certain brain regions, but not the entire brain, can be related to SZ etiology and/or pathophysiology (Garey et al., 1998; Glantz and Lewis, 2000; Kolluri et al., 2005).

Unlike the studies with postmortem brains, in which the results might be affected by long-term medication in the subjects and other confounding factors (Halim et al., 2008; Hashimoto et al., 2007; Li et al., 2004), human brain imaging techniques including fMRI (functional magnetic resonance imaging), PET (positron emission tomography), SPECT (single-photon emission computed tomography), DTI (diffusion tensor imaging), and MRS (magnetic resonance spectroscopy) are able to assess brain functions from first episode-, drug naïve-patients (Davidson and Heinrichs, 2003; McGuire et al., 2008; Pantelis et al., 2003; Sun et al., 2008; Thompson et al., 2001; Vidal et al., 2006; Wright et al., 2000). It has been repeatedly reported that hypofrontality occurs in SZ individuals as evidenced through PET scanning (Andreasen et al., 1992); this was reinforced by the meta-analysis of 41 functional neuroimaging studies (Minzenberg et al., 2009), which suggested that patients with SZ show altered activity with deficits in the dorsolateral prefrontal cortex, anterior cingulate, and mediodorsal nucleus of the thalamus.

Symptomatology has suggested a possibly altered neural circuit for SZ. There is an alteration in sensory motor gating in SZ, which would be regulated by the neural connection among thalamus, limbic cortex, striatum, and pontine tegmentum (Swerdlow et al., 2001). It is now clear that thalamic nuclei, through thalamocortical connections, are key nodes in the establishment of oscillatory dynamics that integrate brain function (Haber and McFarland, 2001; Jones, 2001; Steriade, 2006). There is also a well-documented alteration in working memory and executive function, which implicates the prefrontal cortex-related circuits (Arnsten, 2009; Castner et al., 2004); this is in agreement with a reduction of grey matter volume in the prefrontal cortex in SZ (Wright et al., 2000). The antisaccade task has also been widely employed in SZ research as an oculomotor task, whose deficit is associated with dysfunction of fronto-striatal-thalamo-cortical circuits (Calkins et al., 2007). Taken together, it has been suggested that the abnormality in inter-regional interaction contributes to be the symptoms of SZ, rather than the abnormality of single region (Benes, 2000; Friston, 1998; Garrity et al., 2007) (Fig. 1). Thus, it is crucial to investigate the interaction between brain regions. In this context, gene manipulation by stereotaxic injection into ***multiple*** brain regions could be one of the most ideal approaches for psychiatric search.

Stereotaxic viral injection: actual procedure

Stereotaxic coordinates can be determined from stereotaxic atlases such as *the Rat Brain in Stereotaxic Coordinates* and *the Mouse Brain in Stereotaxic Coordinates*, and coordinates are given as three-dimensional (x, y, and z) distances from an anatomical hallmark such as bregma (Fig. 2). The coordination at x, y, and z axis determines the medial-to-lateral, the anterior-to-posterior, and the dorsal-to-ventral distance from bregma, respectively, which enable precise and reproducible injection into any brain regions in a variety of animals including mouse, rat, songbird, cat, ferret, and even nonhuman primates such as marmoset and macaque monkey. With appropriate anesthesia by use of anesthetics or hypothermia for each developmental stage in an animal subject, animals from a broad developmental range, from postnatal

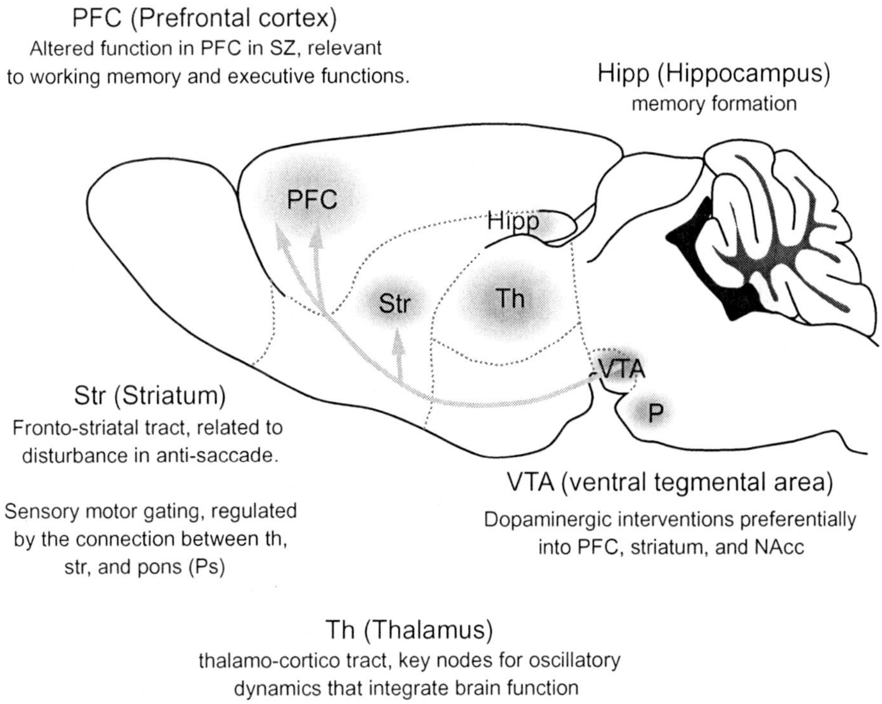

Fig. 1. Examples for altered brain regions presumably relevant to the SZ manifestation. Brain imaging and psychology tests have revealed some manifestations and endophenotypes for SZ. Although neural circuits for rodents and humans are not perfectly comparable anatomically and functionally, this figure depicts conserved brain regions and related neural circuits that are presumably relevant to SZ manifestations.

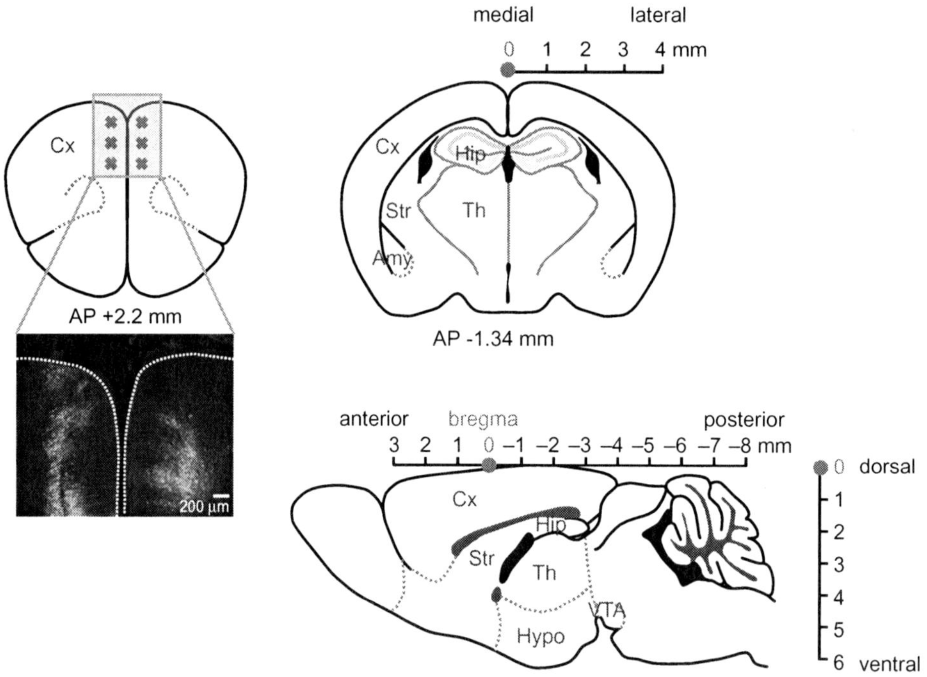

Fig. 2. Stereotaxic coordinates and actual injection of lentivirus-based EGFP. Stereotaxic coordinates were determined from the Rat Brain in Stereotaxic Coordinates. The coordination in Sprague–Dawley rat at postnatal day 21 was stereotaxically injected with EGFP at the coordinates of AP = +2.2; ML = +0.9; DV = +2.0, +1.5, +1.0 from the bregma. The crosses indicate injected sites. Cx, cerebral cortex; Hip, hippocampus; Th, thalamus; Str, striatum; Amy, amygdala; Hypo, hypothalamus; VTA, ventral tegmental area, P, pontine tegmentum. Areas filled with black indicate ventricular regions.

day 3 to adult for rodent, can be studied by this technique (Cetin et al., 2006; Tseng et al., 2008).

Virus vectors for gene manipulation and their properties

With the use of a precise coordination point, any brain region can be functionally altered by coexpression of a heterologous gene or short interfering RNA to induce gene 'knockdown'. As important as the actual injection procedure is the choice of vector to be used for gene delivery. Many viral vectors are available for this purpose, and the selection of vector is determined based on the length of expression desired, the cell preference, the time required to make the vector, and the cytotoxicity, among other considerations (Table 2).

One important factor is the cytotoxicity of the viral vector. For example, Sindbis virus and Semliki Forest virus are both very useful for transient gene expression (Davidson and Breakefield, 2003; Ehrengruber and Lundstrom, 2002), but they are also somewhat cytotoxic, despite improvements in their toxicity (Jeromin et al., 2003; Kim et al., 2004; Lingor et al., 2005). Adenovirus is not cytotoxic itself, but can cause a humoral response if the animal has already been exposed to the adenovirus (Yang et al., 1995).

An obviously important characteristic of a candidate vector is its cell preference; one would like precise cell preference in conjunction with the spatiotemporal precision allowed by stereotaxic injection. Some viruses, such as adeno-associated virus and the alphaviruses (Ehrengruber and Lundstrom, 2007), can be neuron-specific, while others are more general and may affect glia as well. Also, some viruses, such as lentivirus, can have their tropism manipulated using pseudotyping (Cronin et al., 2005). It has additionally been

Table 2. Types of viruses and their properties

Type of virus	Type of experiment	Window of action	Cell preference	Time required to make virus	Cytotoxicity	Infects dividing or nondividing?	Maximum transgene capacity
Adeno-associated virus	Knockdown, gene expression	7 days–6 months	Neuron	Weeks to months	Negligible	Dividing and nondividing	4.5 kb
Adenovirus	Knockdown, gene expression	7 days–6 weeks	Many cell types	Approximately 5 weeks	Negligible	Dividing and nondividing	20 kb
HSV-1	Gene expression	24 h–potentially life long	Neuron	Approximately 2 weeks	Negligible	Dividing and nondividing	50 kb
Lentivirus	Knockdown, gene expression	7 days–3 months	Many cell types	Approximately 1 week	Negligible	Dividing and nondividing	7–8 kb
Retrovirus	Knockdown, gene expression	7 days–6 months	Many cell types	Approximately 2 weeks	Negligible	Dividing	7–8 kb
Semliki Forest virus	Gene expression	24 h–10 days	Neuron	Approximately 1 week	Yes	Dividing and nondividing	6 kb
Sindbis virus	Gene expression	6 h–4 days	Neuron	Approximately 1 week	Yes	Dividing and nondividing	6 kb

demonstrated with lentiviruses that by separating the fusion and binding components on the viral envelope, one can target the virus to specific cells by using antibodies in the envelope that can greatly increase specificity (Yang et al., 2006).

Finally, one must take into account the size of the transgene that one wishes to introduce. HSV-1 has the highest transgene size introduction capability, but this must be weighed against its possible spreading infectivity in the brain (Epstein et al., 2005). Other viral vectors tend to allow introduction of genes between 5 and 10 kb (Davidson and Breakefield, 2003).

Examples of use of stereotaxic injection for psychiatric research

Because of the advantages described above, spatiotemporal gene regulation with use of stereotaxic injection is now becoming utilized for psychiatric research. Mao et al. knocked down the expression of DISC1, one of the promising candidate genes for SZ (Jaaro-Peled et al., 2009), in the adult dentate gyrus by using lentivirus-based DISC1 RNAi, and showed that this injected animal displayed hyperlocomotion in a novel environment, increased immobility in the forced swimming test, but no behavioral changes in the elevated plus maze test (Mao et al., 2009). These suggested that loss of function of one candidate gene for SZ (DISC1) in the adult dentate gyrus plays a role for some, but not all, psychiatric manifestations.

In a *Sapap3*$^{-/-}$ mouse, a model for OCD (obsessive–compulsive disorder), stereotaxic injection of a lentivirus-based overexpression of *Sapap3* into striatum rescues OCD-like symptoms, which clearly suggests the specific significance of Sapap3 function in striatum for the development for OCD (Welch et al., 2007).

It was reported that adult *CD38*$^{-/-}$ mice show marked defects in maternal nurturing because of disruption in oxytocin release via CD38-dependent intracellular Ca^{2+} mobilization. Lentiviral-based delivery of CD38 in the hypothalamus rescues abnormal maternal care in *CD38*$^{-/-}$ mice, which is accompanied with restoration in oxytocin secretion. Taken together, CD38 has a key role in oxytocin release, thereby critically regulating maternal and social behaviors, and that disturbance of oxytocin secretion in hypothalamus may explain some forms of impaired human behavior in the spectrum of autism disorders (Jin et al., 2007).

Stereotaxic injection is also beginning to be used in research on drug, alcohol, and nicotine abuse. Mice with knockout of β2-subunit of the nicotinic acetylcholine receptor have been reported to fail to display nicotine reinforcement behaviors. The reintroduction of the β2-subunit into the VTA (ventral tegmental area) by using a lentiviral vector to reinstigate sensitivity to nicotine reward, however, demonstrates the sufficient role of the VTA in nicotine reinforcement (Maskos et al., 2005).

Conclusions

The advantages of the gene manipulation by the stereotaxic injection technique are (1) excellent spatiotemporal gene expression control in any brain region, (2) the ability to manipulate multiple genes in multiple brain regions simultaneously, (3) cell-type-specific gene manipulation by selection of viral vector and promoter, (4) its simple and easy method. By utilizing this technique, molecular dissection of higher brain functions henceforth becomes accessible to *in vivo* investigation at the cellular and neuronal network level. These advantages will provide crucial clues to understand a complex and genetically polygenic disorders such as SZ.

Acknowledgments

We thank Dr. P. Talalay for manuscript preparation. We appreciate Dr. A. Sawa for scientific discussions. This work was supported by NARSAD (A.H-T) and American Psychiatric Association (AJ. S).

References

Andreasen, N. C., Rezai, K., Alliger, R., Swayze, V. W., 2nd, Flaum, M., Kirchner, P., et al. (1992). Hypofrontality in neuroleptic-naive patients and in patients with chronic schizophrenia. Assessment with xenon 133 single-photon emission computed tomography and the Tower of London. *Archives in General Psychiatry, 49*, 943–958.

Andreasson, S., Allebeck, P., Engstrom, A., & Rydberg, U. (1987). Cannabis and schizophrenia. A longitudinal study of Swedish conscripts. *Lancet, 2*, 1483–1486.

Arnold, S. E., Talbot, K., & Hahn, C. G. (2005). Neurodevelopment, neuroplasticity, and new genes for schizophrenia. *Progress in Brain Research, 147*, 319–345.

Arnsten, A. F. (2009). Stress signalling pathways that impair prefrontal cortex structure and function. *Nature Reviews Neuroscience, 10*, 410–422.

Benes, F. M. (2000). Emerging principles of altered neural circuitry in schizophrenia. *Brain Research Reviews, 31*, 251–269.

Brown, A. S., Begg, M. D., Gravenstein, S., Schaefer, C. A., Wyatt, R. J., Bresnahan, M., et al. (2004). Serologic evidence of prenatal influenza in the etiology of schizophrenia. *Archives in General Psychiatry, 61*, 774–780.

Buka, S. L., Cannon, T. D., Torrey, E. F., & Yolken, R. H. (2008). Maternal exposure to herpes simplex virus and risk of psychosis among adult offspring. *Biological Psychiatry, 63*, 809–815.

Calkins, M. E., Dobie, D. J., Cadenhead, K. S., Olincy, A., Freedman, R., Green, M. F., et al. (2007). The consortium on the genetics of endophenotypes in schizophrenia: Model recruitment, assessment, and endophenotyping methods for a multisite collaboration. *Schizophrenia Bulletin, 33*, 33–48.

Cannon, M., Jones, P. B., & Murray, R. M. (2002). Obstetric complications and schizophrenia: Historical and meta-analytic review. *American Journal of Psychiatry, 159*, 1080–1092.

Cannon, T. D., van Erp, T. G., Bearden, C. E., Loewy, R., Thompson, P., Toga, A. W., et al. (2003). Early and late neurodevelopmental influences in the prodrome to schizophrenia: Contributions of genes, environment, and their interactions. *Schizophrenia Bulletin, 29*, 653–669.

Carter, C. J. (2008). Schizophrenia susceptibility genes directly implicated in the life cycles of pathogens: Cytomegalovirus, influenza, herpes simplex, rubella, and toxoplasma gondii. *Schizophrenia Bulletin*,

Castner, S. A., Goldman-Rakic, P. S., & Williams, G. V. (2004). Animal models of working memory: Insights for targeting cognitive dysfunction in schizophrenia. *Psychopharmacology (Berl), 174*, 111–125.

Cetin, A., Komai, S., Eliava, M., Seeburg, P. H., & Osten, P. (2006). Stereotaxic gene delivery in the rodent brain. *Nature Protocols, 1*, 3166–3173.

Cronin, J., Zhang, X. Y., & Reiser, J. (2005). Altering the tropism of lentiviral vectors through pseudotyping. *Current Gene Therapy, 5*, 387–398.

Davidson, B. L., & Breakefield, X. O. (2003). Viral vectors for gene delivery to the nervous system. *Nature Reviews Neuroscience, 4*, 353–364.

Davidson, L. L., & Heinrichs, R. W. (2003). Quantification of frontal and temporal lobe brain-imaging findings in schizophrenia: A meta-analysis. *Psychiatry Research, 122*, 69–87.

Davis, K. L., Kahn, R. S., Ko, G., & Davidson, M. (1991). Dopamine in schizophrenia: A review and reconceptualization. *American Journal of Psychiatry, 148*, 1474–1486.

Ehrengruber, M. U., & Lundstrom, K. (2002). Semliki Forest virus and Sindbis virus vectors. *Current Protocols in Human Genetics*, Chapter 12, Unit 12. 12.

Ehrengruber, M. U., & Lundstrom, K. (2007). Alphaviruses: Semliki Forest virus and Sindbis virus vectors for gene transfer into neurons. *Current Protocols in Neuroscience*, Chapter 4, Unit 4. 22.

Epstein, A. L., Marconi, P., Argnani, R., & Manservigi, R. (2005). HSV-1-derived recombinant and amplicon vectors for gene transfer and gene therapy. *Current Gene Therapy, 5*, 445–458.

Friston, K. J. (1998). The disconnection hypothesis. *Schizophrenia Research, 30*, 115–125.

Garey, L. J., Ong, W. Y., Patel, T. S., Kanani, M., Davis, A., Mortimer, A. M., et al. (1998). Reduced dendritic spine density on cerebral cortical pyramidal neurons in schizophrenia. *Journal of Neurology, Neurosurgery and Psychiatry, 65*, 446–453.

Garrity, A. G., Pearlson, G. D., McKiernan, K., Lloyd, D., Kiehl, K. A., & Calhoun, V. D. (2007). Aberrant "default mode" functional connectivity in schizophrenia. *American Journal of Psychiatry, 164*, 450–457.

Geyer, M. A. (1998). Behavioral studies of hallucinogenic drugs in animals: Implications for schizophrenia research. *Pharmacopsychiatry, 31*(Suppl 2), 73–79.

Glantz, L. A., & Lewis, D. A. (2000). Decreased dendritic spine density on prefrontal cortical pyramidal neurons in schizophrenia. *Archives in General Psychiatry, 57*, 65–73.

Goff, D. C., & Coyle, J. T. (2001). The emerging role of glutamate in the pathophysiology and treatment of schizophrenia. *American Journal of Psychiatry, 158*, 1367–1377.

Gottesman, I. (1991). *Schizophrenia genesis: The origins of madness*New York: W.H. Freeman.

Haber, S., & McFarland, N. R. (2001). The place of the thalamus in frontal cortical-basal ganglia circuits. *Neuroscientist, 7*, 315–324.

Halim, N. D., Lipska, B. K., Hyde, T. M., Deep-Soboslay, A., Saylor, E. M., Herman, M. M., et al. (2008). Increased lactate levels and reduced pH in postmortem brains of schizophrenics: Medication confounds. *Journal of Neuroscience Methods, 169*, 208–213.

Harrison, P. J. (1999). The neuropathology of schizophrenia. A critical review of the data and their interpretation. *Brain, 122* (Pt 4), 593–624.

Harrison, P. J., & Weinberger, D. R. (2005). Schizophrenia genes, gene expression, and neuropathology: On the matter of their convergence. *Molecular Psychiatry, 10*, 40–68. Image 45.

Hashimoto, K., Sawa, A., & Iyo, M. (2007). Increased levels of glutamate in brains from patients with mood disorders. *Biological Psychiatry, 62*, 1310–1316.

Jaaro-Peled, H., Hayashi-Takagi, A., Seshadri, S., Kamiya, A., Brandon, N.J., & Sawa, A. (2009). Neurodevelopmental mechanisms of schizophrenia: Understanding disturbed postnatal brain maturation through Neuregulin-1 and DISC1. Trends in Neuroscience, *32*(9), 485–495. [Epub 2009 Aug 26.]

Javitt, D. C., & Zukin, S. R. (1991). Recent advances in the phencyclidine model of schizophrenia. *American Journal of Psychiatry*, *148*, 1301–1308.

Jeromin, A., Yuan, L. L., Frick, A., Pfaffinger, P., & Johnston, D. (2003). A modified Sindbis vector for prolonged gene expression in neurons. *Journal of Neurophysiology*, *90*, 2741–2745.

Jin, D., Liu, H. X., Hirai, H., Torashima, T., Nagai, T., Lopatina, O., et al. (2007). CD38 is critical for social behaviour by regulating oxytocin secretion. *Nature*, *446*, 41–45.

Jones, E. G. (2001). The thalamic matrix and thalamocortical synchrony. *Trends in Neuroscience*, *24*, 595–601.

Kellendonk, C., Simpson, E. H., & Kandel, E. R. (2009). Modeling cognitive endophenotypes of schizophrenia in mice. *Trends in Neuroscience*, *32*, 347–358.

Kim, J., Dittgen, T., Nimmerjahn, A., Waters, J., Pawlak, V., Helmchen, F., et al. (2004). Sindbis vector SINrep (nsP2S726): A tool for rapid heterologous expression with attenuated cytotoxicity in neurons. *Journal of Neuroscience Methods*, *133*, 81–90.

Kolluri, N., Sun, Z., Sampson, A. R., & Lewis, D. A. (2005). Lamina-specific reductions in dendritic spine density in the prefrontal cortex of subjects with schizophrenia. *American Journal of Psychiatry*, *162*, 1200–1202.

Lewis, D. A., & Levitt, P. (2002). Schizophrenia as a disorder of neurodevelopment. *Annual Review of Neuroscience*, *25*, 409–432.

Lewis, D. A., & Lieberman, J. A. (2000). Catching up on schizophrenia: Natural history and neurobiology. *Neuron*, *28*, 325–334.

Lewis, G., David, A., Andreasson, S., & Allebeck, P. (1992). Schizophrenia and city life. *Lancet*, *340*, 137–140.

Li, J. Z., Vawter, M. P., Walsh, D. M., Tomita, H., Evans, S. J., Choudary, P. V., et al. (2004). Systematic changes in gene expression in postmortem human brains associated with tissue pH and terminal medical conditions. *Human Molecular Genetics*, *13*, 609–616.

Lingor, P., Scholl, U., Bahr, M., & Kugler, S. (2005). Functional applications of novel Semliki Forest virus vectors are limited by vector toxicity in cultures of primary neurons in vitro and in the substantia nigra in vivo. *Experimental Brain Research*, *161*, 335–342.

Lipska, B. K., Jaskiw, G. E., & Weinberger, D. R. (1993). Postpubertal emergence of hyperresponsiveness to stress and to amphetamine after neonatal excitotoxic hippocampal damage: A potential animal model of schizophrenia. *Neuropsychopharmacology*, *9*, 67–75.

Mao, Y., Ge, X., Frank, C. L., Madison, J. M., Koehler, A. N., Doud, M. K., et al. (2009). Disrupted in schizophrenia 1 regulates neuronal progenitor proliferation via modulation of GSK3beta/beta-catenin signaling. *Cell*, *136*, 1017–1031.

Maskos, U., Molles, B. E., Pons, S., Besson, M., Guiard, B. P., Guilloux, J. P., et al. (2005). Nicotine reinforcement and cognition restored by targeted expression of nicotinic receptors. *Nature*, *436*, 103–107.

McClellan, J. M., Susser, E., & King, M. C. (2007). Schizophrenia: A common disease caused by multiple rare alleles. *British Journal of Psychiatry*, *190*, 194–199.

McGlashan, T. H., & Hoffman, R. E. (2000). Schizophrenia as a disorder of developmentally reduced synaptic connectivity. *Archives in General Psychiatry*, *57*, 637–648.

McGuire, P., Howes, O. D., Stone, J., & Fusar-Poli, P. (2008). Functional neuroimaging in schizophrenia: Diagnosis and drug discovery. *Trends in Pharmacological Sciences*, *29*, 91–98.

Mednick, S. A., Machon, R. A., Huttunen, M. O., & Bonett, D. (1988). Adult schizophrenia following prenatal exposure to an influenza epidemic. *Archives in General Psychiatry*, *45*, 189–192.

Minzenberg, M. J., Laird, A. R., Thelen, S., Carter, C. S., & Glahn, D. C. (2009). Meta-analysis of 41 functional neuroimaging studies of executive function in schizophrenia. *Archives in General Psychiatry*, *66*, 811–822.

Moghaddam, B. (2003). Bringing order to the glutamate chaos in schizophrenia. *Neuron*, *40*, 881–884.

Mortensen, P. B., Pedersen, C. B., Westergaard, T., Wohlfahrt, J., Ewald, H., Mors, O., et al. (1999). Effects of family history and place and season of birth on the risk of schizophrenia. *New England Journal of Medicine*, *340*, 603–608.

Owen, M. J., Craddock, N., & O'Donovan, M. C. (2005). Schizophrenia: Genes at last? *Trends in Genetics*, *21*, 518–525.

Pantelis, C., Velakoulis, D., McGorry, P. D., Wood, S. J., Suckling, J., Phillips, L. J., et al. (2003). Neuroanatomical abnormalities before and after onset of psychosis: A cross-sectional and longitudinal MRI comparison. *Lancet*, *361*, 281–288.

Pedersen, C. B., & Mortensen, P. B. (2001). Evidence of a dose–response relationship between urbanicity during upbringing and schizophrenia risk. *Archives in General Psychiatry*, *58*, 1039–1046.

Pierce, R. C., & Kalivas, P. W. (1997). A circuitry model of the expression of behavioral sensitization to amphetamine-like psychostimulants. *Brain Research Reviews*, *25*, 192–216.

Rapoport, J. L., Addington, A. M., Frangou, S., & Psych, M. R. (2005). The neurodevelopmental model of schizophrenia: Update 2005. *Molecular Psychiatry*, *10*, 434–449.

Sawa, A., & Snyder, S. H. (2002). Schizophrenia: Diverse approaches to a complex disease. *Science*, *296*, 692–695.

Stefansson, H., Ophoff, R. A., Steinberg, S., Andreassen, O.A., Cichon, S., Rujescu, D., et al. (2009). Common variants conferring risk of schizophrenia. *Nature*, *460*, 744–747.

Steriade, M. (2006). Grouping of brain rhythms in corticothalamic systems. *Neuroscience*, *137*, 1087–1106.

Sun, D., Stuart, G. W., Jenkinson, M., Wood, S. J., McGorry, P.D., Velakoulis, D., et al. (2008). Brain surface contraction mapped in first-episode schizophrenia: A longitudinal magnetic resonance imaging study. *Molecular Psychiatry*,

Swerdlow, N. R., Geyer, M. A., & Braff, D. L. (2001). Neural circuit regulation of prepulse inhibition of startle in the rat: Current knowledge and future challenges. *Psychopharmacology (Berl)*, *156*, 194–215.

Tabata, H., & Nakajima, K. (2001). Efficient in utero gene transfer system to the developing mouse brain using electroporation: Visualization of neuronal migration in the developing cortex. *Neuroscience*, *103*, 865–872.

Thompson, P. M., Vidal, C., Giedd, J. N., Gochman, P., Blumenthal, J., Nicolson, R., et al. (2001). Mapping adolescent brain change reveals dynamic wave of accelerated gray matter loss in very early-onset schizophrenia. *Proceedings of the National Academy of Sciences United States of America*, *98*, 11650–11655.

Torrey, E. F., Miller, J., Rawlings, R., & Yolken, R. H. (1997). Seasonality of births in schizophrenia and bipolar disorder: A review of the literature. *Schizophrenia Research*, *28*, 1–38.

Tseng, K. Y., Chambers, R. A., & Lipska, B. K. (2008). The neonatal ventral hippocampal lesion as a heuristic neurodevelopmental model of schizophrenia. *Behavioral Brain Research*,

Tsien, J. Z., Chen, D. F., Gerber, D., Tom, C., Mercer, E. H., Anderson, D. J., et al. (1996). Subregion- and cell type-restricted gene knockout in mouse brain. *Cell*, *87*, 1317–1326.

Vidal, C. N., Rapoport, J. L., Hayashi, K. M., Geaga, J. A., Sui, Y., McLemore, L. E., et al. (2006). Dynamically spreading frontal and cingulate deficits mapped in adolescents with schizophrenia. *Archives in General Psychiatry*, *63*, 25–34.

Weber, K., Rockstroh, B., Borgelt, J., Awiszus, B., Popov, T., Hoffmann, K., et al. (2008). Stress load during childhood affects psychopathology in psychiatric patients. *BMC Psychiatry*, *8*, 63.

Weinberger, D. R. (1987). Implications of normal brain development for the pathogenesis of schizophrenia. *Archives in General Psychiatry*, *44*, 660–669.

Welch, J. M., Lu, J., Rodriguiz, R. M., Trotta, N. C., Peca, J., Ding, J. D., et al. (2007). Cortico-striatal synaptic defects and OCD-like behaviours in Sapap3-mutant mice. *Nature*, *448*, 894–900.

Winterer, G., & Weinberger, D. R. (2004). Genes, dopamine and cortical signal-to-noise ratio in schizophrenia. *Trends in Neuroscience*, *27*, 683–690.

Wright, I. C., Rabe-Hesketh, S., Woodruff, P. W., David, A. S., Murray, R. M., & Bullmore, E. T. (2000). Meta-analysis of regional brain volumes in schizophrenia. *American Journal of Psychiatry*, *157*, 16–25.

Yang, L., Bailey, L., Baltimore, D., & Wang, P. (2006). Targeting lentiviral vectors to specific cell types in vivo. *Proceedings of the National Academy of Sciences United States of America*, *103*, 11479–11484.

Yang, Y., Li, Q., Ertl, H. C., & Wilson, J. M. (1995). Cellular and humoral immune responses to viral antigens create barriers to lung-directed gene therapy with recombinant adenoviruses. *Journal of Virology*, *69*, 2004–2015.

Yolken, R. H., & Torrey, E. F. (1995). Viruses, schizophrenia, and bipolar disorder. *Clinical Microbiology Reviews*, *8*, 131–145.

A. Sawa (Ed.)
Progress in Brain Research, Vol. 179
ISSN 0079-6123

CHAPTER 4

ENU-induced mutant mice for a next-generation gene-targeting system

Yoichi Gondo* and Ryutaro Fukumura

RIKEN BioResource Center, Tsukuba, Ibaraki, Japan

Abstract: By the *N*-ethyl-*N*-nitrosourea (ENU)-based gene-driven mutagenesis, it is now possible to obtain allelic series of mutant mouse strains, each of which carries a different base substitution in any target gene. This new reverse genetic tool has become available based on the ENU mutant mouse library. The ENU mutant mouse library consists of dual archives of frozen sperm and corresponding genomic DNA derived from Generation-1 (G1) male mice, each of which carries thousands of ENU-induced base substitutions. Firstly, ENU-induced mutations in the target gene are screened from the genomic DNA archive by using one of the high-throughput mutation discovery systems. The identified mutations are then revived as live mice by the *in vitro* fertilization (IVF) and embryo transfer (ET) technology. Just like the knockout (KO) mouse system, the revived mutant strains are finally subjected to the three-generation scheme to reveal the gene function(s) of the target gene. This new reverse genetics or "next-generation gene-targeting system" allows us to elucidate the biological roles of the mouse genome in terms of single base-pair effects not only for the protein-coding sequences but also for any genomic sequences.

Keywords: ENU mutagenesis; mutant mouse library; reverse genetics; gene-driven mutagenesis; next-generation gene-targeting; Disc1; Srr

Gene-targeting or knockout (KO) mouse system has been a widely used method to elucidate gene and genome function in the mouse (Thomas et al., 1986; Doetschman, et al., 1988). Because of the evolutionary conservation, KO mice are considered to be a good model for human diseases with respect to gene functions and gene to environmental interactions. The KO mouse is developed by the disruption of the target gene by using the homologous recombination in mouse embryonic stem (ES) cells. Only the KO mouse has been the method to enable the site-directed mutagenesis in the mammalian genome. Live mice carrying the disrupted gene are delivered through the germline chimeras constructed with the manipulated ES cells. Then, the biological role of the gene will be reflected in the phenotype of the KO mice. Thus, it is called gene-driven mutagenesis or reverse genetics.

Quite often, however, the KO mice are resulted in the embryonic lethality due to an essential function(s) in the early development, which hampers the functional studies of the gene in mature living organisms. To bypass the embryonic

*Corresponding author.
Tel.: +81 29 836 9232; Fax: +81 29 836 9098;
E-mail: gondo@brc.riken.jp

DOI: 10.1016/S0079-6123(09)17904-9

lethality, conditional targeting has been facilitated, which allows abolishing the gene in a particular tissue and/or developmental stage (reviewed by Lewandoski, 2001). At the same time, several site-directed mutagenesis method to induce a base substitution in a target base pair has also been developed (Askew et al., 1993; Gondo et al., 1994; Stacey et al., 1994; Wu et al., 1994). These reverse genetic methods have become indispensable tools for the studies of the gene function and for the development of human disease models. In this chapter, another quick-and-thorough gene-driven mutagenesis tool, next-generation gene-targeting system (Gondo, 2008; Gondo et al., 2009), to provide mutant mice carrying various allelic series of point mutations in any target genes and genomic sequences is reviewed. Some examples of *N*-ethyl-*N*-nitrosourea (ENU) mutant strains that provide human disease models in neurological traits are also described.

ENU mouse mutagenesis

ENU is one of the most potent alkylating mutagens and mostly, if not all, induces base substitutions (Justice et al., 1999; Noveroske et al., 2000). ENU may not be used for the site-directed mutagenesis or "gene targeting" because of the random nature of the chemical mutagenesis. Thus, ENU has been used for the phenotype-driven mutagenesis since mid-1990s in the mouse for the genome-wide massive mutant production (e.g., Hrabé de Angelis et al., 2000; Nolan et al., 2000). In spite of the random mutagenesis, we may be able to retrieve ENU-induced point mutations in any target genes when a large number of random mutations enough to cover all genes and genomic sequences are archived. Such possibilities have been extensively examined (Augustin, et al., 2004; Quwailid et al., 2004; Michaud, et al., 2005; Sakuraba, et al., 2005). As depicted in Fig. 1, the key infrastructures that made the ENU-based gene-driven mutagenesis possible in the mouse are (1) the construction of ENU mutant mouse library that is large enough to cover all the mouse genes with multiple ENU-induced point mutation alleles and (2) the development of high-throughput point mutation discovery systems in the target genes from the archive.

ENU mutant mouse library

The ENU mutant mouse library consists of dual archives: frozen sperm and genomic DNA of a large number of G1 mice (Figs. 1c, d).

Frozen sperm archive

All the G1 males were subjected to the cryopreservation of sperm described by Nakagata (2000) and semi-permanently stored in liquid nitrogen tanks. Comparing to the ES cell system to revive the mutant as live mice through germline chimera, the IVF/ET method using the frozen sperm is much simpler and quicker (Fig. 1f). At RIKEN, for instance, approximately 10,000 G1 males were archived as the ENU mutant mouse library (Sakuraba et al., 2005).

Genomic DNA archive

In order to identify ENU-induced point mutations in target genes, PCR primer pairs were firstly designed to amplify the target sequences. Then, the genomic DNA archive was screened by PCR with the high-throughput mutation discovery system described below. The genomic DNA may be isolated from the testis that is surgically excided from each G1 male for the sperm cryopreservation. At present, RIKEN has archived roughly 8000 G1 genomic DNA for the mutation screening.

High-throughput mutation discovery systems

The most critical issue for the ENU-based gene-driven mutagenesis was how to effectively detect the ENU-induced single point mutations in a large number of genomic DNA. The simplest way to detect point mutations would have been the direct sequencing of the PCR products of the target gene; however, it was not realistic in terms of manpower and cost burdens. Thus, much quicker and more cost-effective mutation discovery

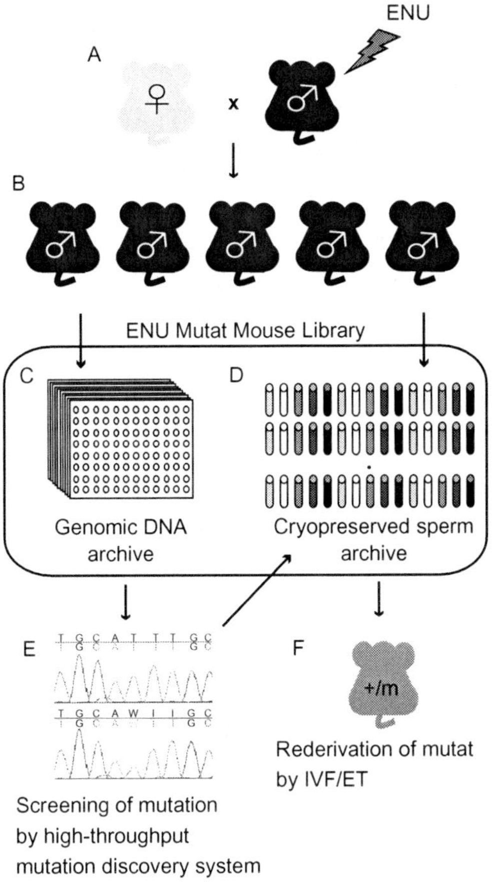

Fig. 1. Schematic flow of ENU-based gene-driven mouse mutagenesis. (A) ENU treated males (G0) were mated to untreated female mice. (B) Each G1 mouse independently inherited ENU-induced mutations from a germ cell of a G0 male. G1 carried all the ENU-induced mutations heterozygously. (C) The genomic DNA were extracted from G1 mice and archived. (D) The G1 sperm samples were also frozen and archived in liquid nitrogen tanks. (E) By using the PCR technology, target sequences in the genomic DNA archive were screened with the high-throughput mutation discovery system. (F) When an ENU-induced mutation was discovered, the corresponding G1 sperm was subjected to the IVF/ET technology to produce G2 mice. Since all the ENU-induced mutations in the G1 mice are heterozygous, half of the G2 mice carry the discovered mutation heterozygously.

method was necessary. Heteroduplex detection methods were chosen for a primary mutation screening, since all the ENU-induced mutations in the G1 genome were heterozygous (see Figs. 1a, b). Temperature gradient capillary electrophoresis (Murphy et al., 2003), Cel1-based mismatch cleavage (Till et al., 2003) and high-resolution melting (Wittwer et al., 2003) have been rapid and sensitive enough for the primary screening. Only the candidate genomic DNA samples were then subjected to the direct sequencing to identify which base pair was mutated in the target gene (Fig. 1e).

Rates and spectrum of ENU-induced mutations

The feasibility of the ENU-based gene-driven mutagenesis was vindicated by the estimation of the ENU-induced mutation rate in the mouse. Summarizing large-scale mutation studies (Augustin et al., 2004; Quwailid et al., 2004; Michaud et al., 2005; Sakuraba et al., 2005), overall ENU-induced mutation rate is estimated to be one in mega base pair of the G1 mouse genome. Considering the conventional protein-coding sequences to be 1000–2000 bp, the mutation rate per protein-coding sequence in the G1 genome is expected to be $1–2 \times 10^{-3}$. It implies that it is possible to find 10–20 ENU-induced point mutations in a particular target-coding sequence by screening entire ENU mutant mouse library consisting of the 10,000 G1 mouse archive at RIKEN. The mutation spectrum of about 100 ENU-induced mutations discovered in various coding sequences (Sakuraba, et al., 2005; Takahasi, et al., 2007) was 60% of missense, 10% of KO-type, and 30% of synonymous mutations (Table 1).

Examples of model mice from the library

The ENU mutant mouse library at RIKEN has been open to research community since 2002. Researchers who are interested in having mutant strains carrying allelic series of ENU-induced point mutations in their target gene may request the screening of the RIKEN library through http://www.brc.riken.go.jp/lab/mutants/genedriven.htm. With respect to schizophrenia, the disrupted in schizophrenia 1 (Disc1) and serine racemase (Srr) genes were screened as candidates based upon researchers' requests and two and one mutant strains were extensively studied by Clapcote et al. (2007) and Labrie et al. (2009), respectively.

Table 1. Spectrum of mutations found in coding sequences of RIKEN ENU mutant mouse library

Type of mutation	Number	Frequency (%)
Missense	56	61.5
Makesense	1	1.1
Nonsense*	3	3.3
Splicing signal*	5	5.5
Synonymous	26	28.6
Total	91	100

"Splicing signals" that are the dinucleotide sequences of splicing donors and acceptors are regarded as coding sequences, because a mutation in such sequences alters the mRNA splicing causing a drastic change of peptide sequence or resulted in nonsense mediated decay. Mutations with asterisks are categorized to the KO-type.

Disc1 gene

DISC1 is one of the risk factors for schizophrenia. It was initially identified at translocation (1;11) (q42.1;q14.3) that was linked to major mental sickness, including schizophrenia, depression, and bipolar disorder in a large Scottish family. Toronto's group designed three pairs of PCR primers, which cover entire exon 2 of the mouse Disc1 gene. We identified four point mutations (Table 2) and Toronto's group decided to analyze two missense mutations, Q31L and L100P. Extensive behavioral, pharmacological, anatomical, and biochemical analyses indicated that Q31L and L100P exhibited recessive depression and dominant schizophrenia models, respectively (Clapcote et al., 2007). The ENU-based gene-driven mutagenesis was shown to have a power to reveal the biological function of the gene at the level of each amino acid residue of the target gene.

Srr gene

Decreased *N*-methyl-D-aspartate receptor (NMDAR)-mediated signaling has been implicated in the pathophysiology of schizophrenia. Activity of NMDAR is modulated by D-serine. Pharmacologically, D-serine also has a similar therapeutic effect for schizophrenia to clozapine. Thus, the SRR gene, which is known to endogenously synthesize D-serine from L-serine, has been a candidate gene as a risk factor for schizophrenia. We screened the RIKEN mutant mouse library with four PCR primer pairs to cover exon 3, 4, 5, 8, and 9 of the mouse Srr gene based upon a request (Table 2). As a result, we found a total of nine point mutations, one of which was a nonsense mutation at the beginning of the last exon (exon 9) of the Srr gene. The behaviors, pharmacological responses, and biochemical profiles of the mice carrying the nonsense mutation in exon 9 of the Srr gene were analyzed in detail (Labrie et al., 2009). The behavioral and pharmacological studies showed the relevant phenotypes for schizophrenia. The Srr mRNA in the homozygotes for the

Table 2. Summary of mutation screening for Discl and Srr genes

Amplicon name	Target length(bp)[a]	ORF (bp)	Analyzed number of G1 mice	Sum of analyzed target length (bp)	Number of identified mutation
Disc1-EX2-1	501	396	1,689	846,189	0
Disc1-EX2-2	563	490	1,682	946,966	3
Disc1-EX2-3	357	357	4,494	1,604,358	1
Srr-EX3	494	168	7,460	3,685,240	1
Srr-EX4, 5	437	164	7,463	3,261,331	4
Srr-EX8	376	210	1,689	635,064	0
Srr-EX9	513	215	7,401	3,7967,13	4

[a]Target length is the size of the PCR product from which the primer lengths are subtracted.

nonsense mutation was reduced to be 50% of the wild-type level. The Srr protein was totally abolished in the homozygotes, indicating the KO-equivalent null allele of the nonsense mutation. Indeed, Basu et al. (2009) independently gene-targeted and studied the KO mice for the Srr gene and found equivalent results; thus, the ENU-based gene-driven mutagenesis also provides KO-type alleles in addition to various allelic series of missense mutations.

Future perspectives

Three mutations in coding sequences were described in this chapter as examples of mouse models for human diseases derived from the ENU mutant mouse library. The library, however, encompasses not only coding mutations but also many point mutations in the mouse genome, most of which is noncoding sequences. The RIKEN library, which consisted of 10,000 of G1 mice, has 3×10^7 ENU-induced mutations in the mouse genome, giving one mutation every 100 bp on an average (Gondo, 2008; Gondo et al., 2009). Currently, mutations have been screened based upon researchers' request. In near future, the next-generation ultra-throughput re-sequencer system should identify all the point mutations in the ENU mutant mouse library, making various allelic series of mutant mouse strains immediately available to the research community.

Acknowledgments

This work is partly supported by a Grant-in-Aid for Scientific Research (A) to Y. G. (KAKENHI 21240043), for Grants-in-Aid for Young Scientists (B) to R. F. (KAKENHI 20790196) from Japan Ministry of Education, Science, Sports and Culture (MEXT) and Japan Society for the Promotion of Science (JSPS).

List of Abbreviations:

ENU	*N*-ethyl-*N*-nitrosourea
G0, G1, G2	Generation-0, Generation-1, Generation-2
IVF	*in vitro* fertilization
ET	embryo transfer
KO mouse	knockout mouse
ES	embryonic stem
DISC1, Disc1	disrupted in schizophrenia 1
SRR, Srr	serine racemase
NMDAR	*N*-methly-D-aspartate receptor

References

Askew, G. R., Doetschman, T., & Lingrel, J. B. (1993). Site-directed point mutations in embryonic stem cells: A gene-targeting tag-and-exchange strategy. *Molecular Cell Biology*, *13*(no. 7), 4115–4124.

Augustin, M., Sedlmeier, R., Peters, T., Huffstadt, U., Kochmann, E., Simon, D., et al. (2004). Efficient and fast targeted production of murine models based on ENU mutagenesis. *Mammalian Genome*, *16*(no. 6), 405–413.

Basu, A. C., Tsai, G. E., Ma, C. L., Ehmsen, J. T., Mustafa, A. K., Han, L., et al. (2009). Targeted disruption of serine racemase affects glutamatergic neurotransmission and behavior. *Molecular Psychiatry*, *14*(no. 7), 719–727.

Clapcote, S. J., Lipina, T. V., Millar, J. K., Mackie, S., Christie, S., Ogawa, F., et al. (2007). Behavioral phenotypes of Disc1 missense mutations in mice.. *Neuron*, *54*(no. 3), 387–402.

Doetschman, T., Maeda, N., & Smithies, O. (1988). Targeted mutation of the Hprt gene in mouse embryonic stem cells. *Proceedings of the National Academy of Sciences United States of America*, *85*(no. 22), 8583–8587.

Gondo, Y. (2008). Trends in large-scale mouse mutagenesis: From genetics to functional genomics. *Nature Reviews Genetics*, *9*(no. 10), 803–810.

Gondo, Y., Nakamura, K., Nakao, K., Sasaoka, T., Ito, K., Kimura, M., et al. (1994). Gene replacement of the p53 gene with the lacZ gene in mouse embryonic stem cells and mice by using two steps of homologous recombination. *Biochemical and Biophysical Research Communications*, *202*(no. 2), 830–837.

Gondo, Y., Fukumura, R., Murata, T., & Makino, S. (2009). Next-generation gene targeting in the mouse for functional genomics. *Biochemistry and Molecular Biology Reports*, *42* (no. 6), 315–323.

Hrabé de Angelis, M. H., Flaswinkel, H., Fuchs, H., Rathkolb, B., Soewarto, D., Marschall, S., et al. (2000). Genome-wide, large scale production of mutant mice by ENU mutagenesis. *Nature Genetics*, *25*(no. 4), 444–447.

Justice, M. J., Noveroske, J. K., Weber, J. S., Zheng, B., & Bradley, A. (1999). Mouse ENU mutagenesis. *Human Molecular Genetics*, *8*(no. 10), 1955–1963.

Labrie, V., Fukumura, R., Rastogi, A., Fick, L. J., Wang, W., Boutros, P. C., et al. (2009). Serine racemase is associated with schizophrenia susceptibility in humans and in a mouse model. *Human Molecular Genetics*,

Lewandoski, M. (2001). Conditional control of gene expression in the mouse. *Nature Reviews Genetics*, *2*(no. 10), 743–755.

Michaud, E. J., Culiat, C. T., Klebig, M. L., Barker, P. E., Cain, K. T., Carpenter, D. J., et al. (2005). Efficient gene-driven germ-line point mutagenesis of C57BL/6J mice. *BMC Genomics*, *6*, 164.

Murphy, K., Hafez, M., Philips, J., Yarnell, K., Gutshall, K., et al. (2003). Evaluation of temperature gradient capillary electrophoresis for detection of the factor v leiden mutation: Coincident identification of a novel polymorphism in factor v. *Molecular Diagnostics*, *7*(no. 1), 35–40.

Nakagata, N. (2000). Cryopreservation of mouse spermatozoa. *Mammalian Genome*, *11*(no. 7), 572–576.

Nolan, P. M., Peters, J., Strivens, M., Rogers, D., Hagan, J., Spurr, N., et al. (2000). A systematic, genome-wide, phenotype-driven mutagenesis programme for gene function studies in the mouse. *Nature Genetics*, *25*(no. 4), 440–443.

Noveroske, J. K., Weber, J. S., & Justice, M. J. (2000). The mutagenic action of *N*-ethyl-*N*-nitrosourea in the mouse. *Mammalian Genome*, *11*(no. 7), 478–483.

Quwailid, M. M., Hugill, A., Dear, N., Vizor, L., Wells, S., Horner, E., et al. (2004). A gene-driven ENU-based approach to generating an allelic series in any gene. *Mammalian Genome*, *15*(no. 8), 585–591.

Sakuraba, Y., Sezutsu, H., Takahasi, K. R., Tsuchihashi, K., Ichikawa, R., Fujimoto, N., et al. (2005). Molecular characterization of ENU mouse mutagenesis and archives. *Biochemical and Biophysical Research Communications*, *336* (no. 2), 609–616.

Stacey, A., Schnieke, A., McWhir, J., Cooper, J., Colman, A., & Melton, D. W. (1994). Use of double-replacement gene targeting to replace the murine alpha-lactalbumin gene with its human counterpart in embryonic stem cells and mice. *Molecular Cell Biology*, *14*(no. 2), 1009–1016.

Takahasi, K. R., Sakuraba, Y., & Gondo, Y. (2007). Mutational pattern and frequency of induced nucleotide changes in mouse ENU mutagenesis. *BMC Molecular Biology*, *8*, 52.

Thomas, K. R., Folger, K. R., & Capecchi, M. R. (1986). High frequency targeting of genes to specific sites in the mammalian genome. *Cell*, *44*(no. 3), 419–428.

Till, B. J., Reynolds, S. H., Greene, E. A., Codomo, C. A., Enns, L. C., Johnson, J. E., et al. (2003). Large-scale discovery of induced point mutations with high-throughput TILLING. *Genome Research*, *13*(no. 3), 524–530.

Wittwer, C. T., Reed, G. H., Gundry, C. N., Vandersteen, J. G., & Pryor, R. J. (2003). High-resolution genotyping by amplicon melting analysis using LCGreen. *Clinical Chemistry*, *49*(no. 6), 853–860.

Wu, H., Liu, X., & Jaenisch, R. (1994). Double replacement: Strategy for efficient introduction of subtle mutations into the murine Col1a-1 gene by homologous recombination in embryonic stem cells. *Proceedings of the National Academy of Sciences United States of America*, *91*(no. 7), 2819–2823.

A. Sawa (Ed.)
Progress in Brain Research, Vol. 179
ISSN 0079-6123

CHAPTER 5

Inducible and conditional transgenic mouse models of schizophrenia

Mikhail V. Pletnikov*

Departments of Psychiatry and Neuroscience, Johns Hopkins University School of Medicine, Baltimore, MD, USA

Abstract: Schizophrenia is a devastating disorder. Despite the advance in research techniques in the last couple of decades, the pathogenesis of the disorder still remains poorly understood. Given the lack of pathognomonic feature of the disease and difficulty to analyze molecular pathways in patients, animal models have been instrumental in advancing our understanding of the disease. Recent progress in genetics has indentified candidate susceptibility genes for schizophrenia, and generation of new genetic animal models has begun to provide valuable insights into the disease development. However, the complex neurodevelopmental and heterogeneous nature of schizophrenia still poses tremendous challenges for creating credible mouse models. In this review, we will discuss how current genetic systems of temporal and conditional regulation of gene expression have shed lights on the functions of the candidate genes in mouse models of schizophrenia. We also consider the strength and weaknesses of each model. We will argue that further development of more sophisticated genetic animal models is crucial for clarifying the unknowns of schizophrenia.

Keywords: schizophrenia; Tet-off system; genetics; animal model; DISC1; dopamine; cAMP

Introduction

Schizophrenia is a devastating disorder with a worldwide prevalence of 0.5–1.2% (Kessler et al., 2005; Wu et al., 2006). The disease generally presents at late adolescence and early adulthood (Kessler et al., 2007). Being a disorder of youth and adulthood, schizophrenia is associated with significant loss of daily activity, reduced productivity at work and increased unemployment rates (Wu et al., 2005). Although the efforts to treat the disease have produced the significant success, the current treatment options are mostly palliative (Lieberman, 2005).

For this chronic, debilitating disease which is known for thousands of years (Stone 2006), there is not any test that can correctly identify the affected individuals even after their death. The data obtained from postmortem brain samples have demonstrated the subtle pathology. Understanding the underlying biology of the disease is difficult as postmortem brain samples and clinical assays have not yet identified specific morphological, biochemical or electrophysiological phenotype.

*Corresponding author.
Tel.: +1 410 502 3760; Fax: +1 410 614 0013;
E-mail: mpletnik@jhmi.edu

DOI: 10.1016/S0079-6123(09)17905-0

The genetic epidemiologic studies have suggested that schizophrenia is a multi-factorial disorder with a strong genetic component (Sullivan, 2003). The linkage and association studies have identified several promising candidate genes. Among those, Dysbindin 1 (DTNBP1), neuregulin1 (NRG1), and Disrupted-in-Schizophrenia-1 (DISC1) are examples of candidate genes that are likely involved in the mechanisms of abnormal neurodevelopment, providing a strong genetic support for the neurodevelopmental hypothesis of schizophrenia (Ross et al., 2006; Straub and Weinberger, 2006).

Schizophrenia is thought to arise in part from subtle defects in development of the cerebral cortex, hippocampus, and other forebrain structures (Weinberger, 1996; Lewis and Levitt 2002; Eastwood, 2004; Arnold et al., 2005). The concept of a neurodevelopmental origin of schizophrenia has received wide acceptance, although the evidence so far has been primarily indirect (Marenco and Weinberger 2000). In addition, the term "neurodevelopmental disorder" should be probably extended beyond a single insult during prenatal period to include postnatal and early adulthood abnormalities in fine pruning of dendritic tree, late stages of synaptogenesis and plasticity mechanisms (Woods, 1998). In this context, understanding the developmental nature of psychiatric conditions becomes a crucial task. Elucidating when and where genetic mutations have most profound impacts on neurodevelopment will undoubtedly advance our understanding of the complex mechanisms of schizophrenia.

Thus, mouse models that allow for elucidating timing of effects of the genetic mutations and identification of sensitive periods of prenatal and postnatal abnormal brain and behavior maturation have become important tools in basic research in schizophrenia. In this review, I will focus on inducible and conditional transgenic mouse models of schizophrenia. There are several excellent expert reviews on inducible gene expression systems (e.g., Gingrich et al., 1998, Zhu et al., 2002; Stieger et al., 2009). I will briefly describe the main genetic principles of inducible systems used to generate mouse models and will critically overview existing inducible transgenic models of schizophrenia. Future research in genetic mouse models of schizophrenia will also be discussed.

Inducible gene expression models

Precise control of gene expression is an invaluable tool in studying complex physiological processes. The usefulness of tissue- or cell-specific promoters to express target genes in the selected tissues or cells has been widely used with success (Zhu et al., 2002). However, if the activity of a promoter is constitutive, there is no control over the timing of the expression. In the beginning, a major reason to have a temporal control over-expression of a transgene was to avoid its possible adverse or even lethal developmental effects. This has been, for example, critical for animal models of neurodegenerative diseases as the effects of the mutation are thought to take place mostly in adulthood. Thus, attempts have been made to generate inducible mouse models to avoid potential toxic effects of transgenes during early development (e.g., Yamamoto et al., 2000). In the field of genetic animal models of schizophrenia and related developmental disorders, the issue of temporal regulation has the additional significances. There are always questions whether the resultant phenotype is related to possible compensatory adaptations during development and at what age the effects of the transgene are most profound and lasting.

To address the limitations of the constitutive over-expression system, conditional and inducible gene expression systems have been developed. Several different inducible systems are currently in use (Gingrich et al., 1998; Mallo, 2006). Although the components and structure of each system can be different, the principles of these systems are similar. For an inducible transgenic system to work properly, three parts are needed, a regulatory unit, a responsive element linked to a target gene, and an induction agent. In an ideal system, expression of the transgene can be turned on or off quickly and reversibly by an external inducer. Combined with a tissue- or cell-specific promoter, the inducible system provides temporal and spatial control over the expression of the transgene. The system has no expression in the "Off" state but high levels of expression when the gene is "On". An ideal inducer should be inexpensive and readily available (Zhu et al., 2002). It should be easy to administer the compound in the drinking water or food. It is highly desirable that the compounds have high bioavailability, reach all tissues easily, and pass through the placenta and/or the blood–brain barrier. The agent should have no side effects that interfere with the target phenotype (Zhu et al., 2002).

Several inducible systems are currently available and reviewed elsewhere (Mallo, 2006; Stieger et al., 2009). Here, I will just briefly describe the principles of the two systems that have been used to generate mouse models of schizophrenia, one is based on the tetracycline-controlled transcriptional regulator developed by Gossen et al. (1992, 1995), and the other is based on the ligand-binding domain (LBD) of a human estrogen receptor (Logie and Stewart 1995; Metzger et al., 1995).

The tetracycline transactivator system

The *tet* system, based on the regulatory elements of the *Escherichia coli* tetracycline resistance operon (Hillen et al., 1982; Hillen and Berens, 1994), is the most popular system to control gene expression. In bacteria, transcription of the genes mediates resistance to tetracycline and is under the control of the protein TetR (Tet repressor) that binds to the *tetO* (operator) within the operon promoter to block transcription (Hillen et al., 1982; Hillen and Berens, 1994). The system that is widely used in mouse genetic models is bi-transgenic model that includes two genetic cassettes (Fig. 1). The so-called response cassette consists of the target gene linked to a hybrid promoter that contains tandem repeats of *tetO* and a minimal promoter to express the gene. Transcription of the target gene is controlled by tetracycline or its analogue, doxycycline (DOX) (Degenkolb et al., 1991). The most popular versions of the *tet* system employ a TetR derivative that has been modified to serve as a transcriptional activator by fusing TetR with the herpes simplex virus protein VP16 (16) (Fig. 1). This chimeric protein, known as tTA (tet Transcriptional Activator), binds to the tetO tandem repeats (tetracycline response elements) and activates transcription of the downstream transgene. If DOX is present, it binds to tTA, changes its conformation, and prevents its binding to tetracycline responsive element (TRE), leading to cessation of transcription. In this combination, the system is known as "Tet-off". Certain mutations in the TetR moiety of tTA resulted in a molecule

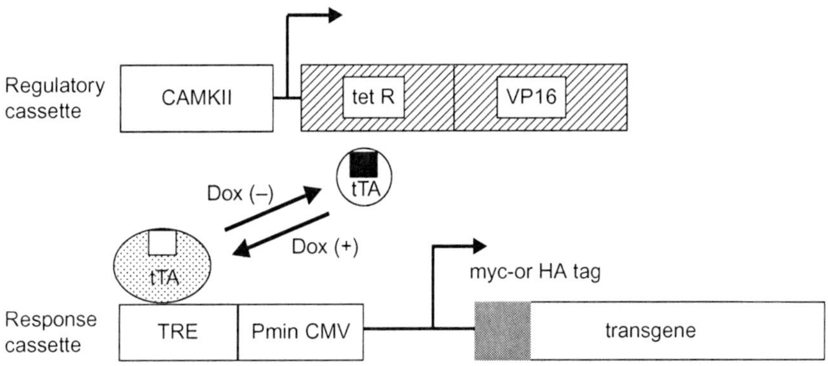

Fig. 1. Schematic representation of the Tet-off system. A Tet system is composed of two distinct expression cassettes, one containing the regulatory protein (tTA) under the control of a cell- type-specific promoter (e.g., the CAMKII promoter) and the second response cassette containing the TRE fused to a minimal promoter (e.g., CMV), regulating the expression of the transgene that is usually fused to a tag for easy detection of transgene expression. The chimeric protein tTA consists of the TetR fused to the viral transactivator VP16. In the absence of Dox, the tTA protein binds to the TRE within the compound promoter P_{min}CMV to allow the expression of the transgene. In the presence of Dox, the tTA changes its conformation, gets detached from the TRE and no transgene expression is observed.

(rtTA) with reversed regulation by DOX that now activates transcription. The *tet* system using rtTA is known as "Tet-on" (Gossen et al., 1995).

In order to generate mouse model with inducible expression of the transgene in living animals, two single transgenic lines are mated, one containing the target gene downstream of the minimal promoter, and another expressing either tTA or rtTA under the control of a suitable promoter (Fig. 1). While temporal regulation of expression of the transgene in double transgenic lines is achieved by adding or withdrawing DOX-containing food or water, the tissue- or cell-specific expression is regulated by the promoter controlling expression of tTA or rtTA. Thus, this system provides both temporal and conditional expression of the transgene.

The tetracycline transactivator system has several distinct advantages over other inducible systems developed for use in transgenic mice. Because the tetracycline system is based on prokaryotic elements, their presence produces minimal pleiotropic effects. The very low or non-existing basal levels of expression that is tightly regulated and can be easily and relatively quickly (days) induced are attractive features of this system. In addition, the well-characterized pharmacological properties (including rapid uptake into cells and penetration through the placenta and the blood/brain barrier) and low toxicity of DOX have facilitated generation of transgenic animals using the Tet system (Kistner et al., 1996; Gingrich et al., 1998).

The mutant estrogen receptor system

The first report about this model has described inducible site-directed recombination in mouse embryonic stem cells where protein consisting of Cre recombinase was fused to a mutated hormone-binding domain of the murine estrogen receptor (Zhang et al., 1996). Conditional Cre activity was achieved by expressing the recombinase as fusion proteins with steroid hormone receptor LBDs (Logie and Stewart, 1995; Metzger et al., 1995). Although the system worked efficiently in vitro, endogenous estrogen present in animals prevented direct application of the system in vivo (Gingrich et al., 1998). It was suggested that use of mutated LBDs, which would selectively respond to synthetic ligands, e.g., tamoxifen, could resolve the problem (Gingrich et al., 1998). One type of this inducible model has cDNA linked to LBD and a site-specific promoter located upstream of the transgene. In the absence of tamoxifen, the synthesized protein is usually sequestered by heat-shock chaperone proteins and degraded (Kida et al., 2002; Li et al., 2007). When tamoxifen is administered to transgenic mice, it binds the LBD and the fusion protein complex undergoes a conformational change, leading to dissociation of the transgenic protein from the chaperone proteins complex. The protein becomes functional. After a quick metabolism of tamoxifen, the protein acquires non-functional state again (Kida et al., 2002; Li et al., 2007). Any potential side effects of tamoxifen are expected to be mild and should rapidly disappear upon withdrawal of the drug.

Inducible genetic mouse models of schizophrenia

Progress in human genetics has identified several candidate genes and, thus, facilitated generation of novel mouse models. While most mouse models based on constitutive over-expression of a gene or its full or partial deletion (i.e., knockout mice) (e.g., Mohn et al., 1999; Hikida et al., 2007; Shen et al., 2008), several inducible and conditional transgenic models of schizophrenia have been generated based on the Tet-off or mutant estrogen receptor systems.

Inducible expression of mutant DISC1

We generated a mouse model of conditional and inducible expression of human mutant DISC1 using the Tet-off system (Pletnikov et al., 2008). Mutant DISC1 is a hypothetical protein product of the balanced t(1;11) chromosomal translocation identified in a Scottish pedigree with high load of major mental disorders, including schizophrenia and major depression (Millar et al., 2001; Ishizuka et al., 2006; Chubb et al., 2008). Fine mapping and cloning have identified a disrupted gene on chromosome 1, hence the name DISC1. As the breakpoint is in the middle of open reading frame, the translocation is hypothesized to produce the truncated N-terminus product, mutant DISC1 (Millar et al., 2001). The identifiable mutation that is strongly associated with major mental diseases makes DISC1 and the mutant protein product interesting and attractive candidates for studying the neurobiology of psychiatric disorders (Ross et al., 2006). There are several examples of how similar functional mutations have helped to shed light on the molecular mechanisms of neurodegenerative diseases, including familial forms of Parkinson's disease and Alzheimer's disease (e.g., Davidzon et al., 2006; Piscopo et al., 2008).

Recent studies have implicated DISC1 in neuronal development, neuronal migration, and synaptogenesis (Ishizuko et al., 2006; Ross et al., 2006; Camargo et al., 2007). They have also suggested that mutant DISC1 may interfere with the functions of normal wild-type (WT) DISC1 via dominant-negative mechanisms, leading to loss-of-function of DISC1 (Kamiya et al., 2005). Thus, we generated transgenic mouse model of inducible and conditional expression of mutant human DISC1 to study the molecular mechanisms whereby this protein affects neurodevelopment.

Our inducible DISC1 mouse model is a standard bi-transgenic Tet-off system (Fig. 1). In order to turn off transgene expression, DOX is added to mouse food or drinking water. As transcription of tTA is regulated by the α-calmodulin kinase II (CAMKII) promoter, expression of mutant DISC1 is present in neurons of the olfactory bulbs, cortex, hippocampus, striatum but not cerebellum. It was found that expression of mutant DISC1 starts prenatally as early as embryonic day (E) 15 as detected by western blot and E9 when assayed by RT-PCR (unpublished data). Thus, our model provides the opportunity to regulate both prenatal and postnatal expression of mutant DISC1.

The initial characterization of our model has included evaluation of the neurobehavioral effects of mutant DISC1 when its expression was present throughout the entire life span of mice. Expression of mutant DISC1 was on the mixed SJL;B6; CBA background (Pletnikov et al., 2008). We found that expression of mutant DISC1 was associated with increased spontaneous locomotor activity in male but not female mice, decreased social interaction and increased aggressive behavior in male mice when measured in open field test, and decreased spatial recognition memory in Morris water maze in female mice only despite comparable rates of learning between mutant and control mice. These alterations are reminiscent of positive and negative symptoms, and cognitive impairments seen in schizophrenia (Ross et al., 2006). No effects of mutant DISC1 were found in pre-pulse inhibition (PPI) of the acoustic startle and novelty-induced activity in open field.

These behavioral alterations were accompanied by enlargement of the lateral ventricle, the most consistent structural pathology seen in schizophrenic patients (Vita et al., 2006; Pagsberg et al., 2007). Ventricular enlargement can be partly explained by attenuated dendritic arborization found in primary cortical neurons derived from mutant DISC1 embryos, in line with human postmortem studies, showing decreased dendritic length and dendritic arborization in certain cortical areas (Glantz et al., 2000). Our biochemical assays showed that the effects of mutant DISC1 may be mediated by binding mutant DISC1 to endogenous mouse Disc1, producing decreased levels of mouse Disc1 and its interacting partner, Lis1, which have been implicated in the molecular mechanisms of neuronal maturation (Morris et al., 2003; Ozeki et al., 2003).

The main drawback of the study is that mutant DISC1 was expressed steadily throughout the entire life. Thus, the contribution of prenatal vs. postnatal periods remained unclear. Our recent experiments with regulation of expression of mutant DISC1 have demonstrated that prenatal and postnatal expression selectively affected different neurobehavioral phenotypes, suggesting the effects of mutant DISC1 may vary across neurodevelopment (manuscript in revision).

Inducible expression of the fragment of human DISC1

The group of investigators led by T. Cannon and A. Silva has recently described an interesting mouse model of inducible expression of the fragment of DISC1 (DISC1-cc) that is deleted in the affected members of the Scottish family. DISC1-cc expression is regulated by the CAMKII promoter, which is active in forebrain neurons. DISC1-cc spans residues 671–852, which is the portion of DISC1 crucial for binding to NUDEL and Lis1 (Morris et al., 2003; Ozeki et al., 2003; Kamiya et al., 2006). The DISC1-cc protein was fused to a HA-tagged mutant (G521R) estrogen receptor LBD (Kida et al., 2002). In this inducible system, the transgenic protein is degraded after sequestration by heat-shock chaperone proteins. When tamoxifen is administered to transgenic mice, it binds the LBD, the fusion protein complex, which includes DISC1-cc, undergoes a conformational switch such that the transgenic protein dissociates from the chaperone proteins and becomes functional. After a quick metabolism of tamoxifen, the transgenic protein acquires non-functional state again (Li et al., 2007).

Consistent with previous reports (Morris et al., 2003; Ozeki et al., 2003; Brandon et al., 2004; Kamiya et al., 2006), DISC1-cc was found to bind to Nudel and Lis1 6 h after induction, but not 2 days after induction. Similar to the study by Pletnikov et al., binding of DISC1-cc to Nudel and Lis1 decreased endogenous DISC1 protein in DISC1/Nudel complexes (dominant-negative effects), indicating that this inducible transgenic system is useful for elucidating the neurobehavioral effects of disruption of endogenous DISC1 function during selective periods of brain development. Expression of the DISC1-cc protein was detected in the cortex, hippocampus, striatum, and cerebellum of the transgenic mice. Transgenic mice appeared normal and displayed no gross abnormalities. Locomotor activity in open field and anxiety-related responses in elevated plus maze were comparable between transgenic and control mice.

By regulating expression of DISC1-cc, the neurobehavioral effects of the dominant-negative DISC1-cc construct were compared between mice that expressed DISC1-cc at postnatal day 7 vs. adulthood. DISC1-cc mice were tested in a spatial working memory test [the delayed non-matched to place (DNMTP) task]. The adult DISC1-cc transgenic mice with induction at postnatal day 7 showed decreased percentage of correct DNMTP choices compared with adult WT mice treated with tamoxifen, adult DISC1-cc transgenic or adult WT mice treated with vehicle. Importantly, no differences between DISC1-cc and WT mice treated with tamoxifen were found when this compound was injected in adulthood. The results clearly showed that disruption of DISC1 function early in development but not adulthood was responsible for deficient spatial working memory.

Another example of the developmental effects of DISC1-cc included depression-like behavior in the forced swimming test. The transgenic mice treated with the tamoxifen at day 7 had shorter latencies to floating compared control groups. In contrast, induction of expression during adulthood had no effects in this test. Thus, similar to the cognitive effects, disruption of endogenous mouse DISC1 early in development but not adulthood produced depressive-like phenotype. The effects of DISC1-cc on social behaviors were assessed in a three-chambered apparatus, which allows for evaluation of sociability and preference for social novelty (Moy et al., 2004). Induction of expression of DISC1-cc at day 7 altered the normal pattern of social activity so that transgenic mice spent the similar amounts of time with a live mouse and an inanimate object, whereas control mice demonstrated a clear preference for the live partner. Thus, social activity was also affected by induction of the fragment early in development, and no such effects were noted because of adult expression of DISC1-cc. Possible alterations in neuronal structure and function during development were analyzed to explain the behavioral deficits. Induction of expression of the fragment at day 7 but not in adulthood produced attenuated dendritic complexity in mice. Notably, this reduction in dendritic arborization was associated with reduced basal synaptic transmission, although presynaptic functions seemed to remain unaffected by activation of the DISC1-cc protein.

Taken together, this study was the first to convincingly demonstrate that disruption of DISC1 functions during early postnatal period but not adulthood can produce a set of behavioral, neuroanatomical, and neurophysiological alterations that are consistent with aspects of schizophrenia and mood disorders. The weak points of the study include expression of the DISC1-cc fragment that is the deleted C-terminal portion of human DISC1, a brief expression of the fragment, and manipulations with postnatal expression only.

Inducible expression of the G-protein subunit gas

Another successful example of using inducible genetic mouse model to explore the developmental etiology of schizophrenia-like neurobehavioral abnormalities has been reported by Kelly and associates (Kelly et al., 2009). Their study sought to evaluate a role of the G-protein subunit Gαs in the pathophysiology of neurobehavioral schizophrenia-like abnormalities in mice. Recently, a polymorphism that increases mRNA levels of Gαs was associated with schizophrenia (Kelly et al., 2009).

The bi-transgenic model used is based on the Tet-off system in which the Gαs transgene with a hemagglutinin epitope that is placed downstream the tetracycline operator (tetO). Gαs transgenic mice was found to express high levels of the transgene restricted to the cortex, hippocampus, dorsal striatum and ventral striatum, and low levels in the amygdala and in the inferior colliculus. Expectedly, expression of the transgene produced increased levels of Gαs mRNA in those regions. As a result, Gαs mice had elevated basal and GTPγs-stimulated adenylate cyclase activity.

The authors analyzed the relative contributions of developmental vs. postnatal expression of the transgene to schizophrenia-related behavioral alterations. Gαs mice exhibited unaltered startle responses, but their PPI was significantly attenuated. Notably, both Gαsdev (developmental expression) and Gαsadult (adult expression only) mice had similar PPI compared to control littermates, suggesting both developmental and adulthood over-expression of Gαs were critical for manifestation of the PPI deficit. Administration of a typical antipsychotic, haloperidol (1 mg/kg, i.p.) was able to ameliorate the PPI deficits in Gαs mice. Compared to control animals, Gαs mice exhibited increased horizontal locomotor activity and a trend toward increased rearing activity in open field. Interestingly, Gαsadult mice, but not Gαsdev mice, demonstrated increased horizontal movements, suggesting that expression of the transgene during adulthood produced locomotor hyperactivity.

Developmental expression of Gαs was necessary and sufficient to produce spatial learning deficits in adult mice. When hippocampus-dependent spatial learning and memory in the hidden-platform Morris water maze were assayed, both Gαs mice (expression throughout development and adulthood) and Gαsdev (developmental expression only) mice displayed significantly greater escape latencies during training compared to control littermates, indicating impaired spatial learning. In contrast, Gαsadult mice with expression during adulthood only did not differ from Gαsdox mice treated with DOX (no expression) in learning rate in the test. Poorer memory of the location of the hidden platform was also detected in Gαs and Gαsdev mice. Gαsadult mice (adult expression only) displayed an intermediate phenotype, while Gαsdox mice (no expression) had no alterations in this test, suggesting that the cognitive effects observed in Gαs mice were likely related to expression of the transgene rather than non-specific gene insertion effects.

As Gαsadult mice did not have any changes in spatial learning but were inferior in the impaired spatial memory phase of MWM, the authors hypothesized that adult over-expression of Gαs might selectively affect memory retrieval. To address this hypothesis, a one-trial Pavlovian fear conditioning task was used. Gαs and control mice showed comparable amounts of freezing during the training session, suggesting that over-expression of the transgene did not alter the pain sensitivity in mice. Cue-dependent short-term and long-term conditional fear was not changed in transgenic mice compared to control littermates. In contrast, Gαs mice displayed significantly impaired short-term and long-term contextual fear that is hippocampus-dependent. When expression of the transgene was present only during development, no abnormalities were seen in long-term memory for contextual fear. However, adult expression of the transgene was associated with lower levels of contextual fear.

Using an elegant experimental design, the authors tried to evaluate if the effects of Gαs on contextual fear were related to selective alterations in acquisition, consolidation or retrieval. They compared control littermates to Gαs mice that expressed the transgene during training but not retrieval by administering DOX beginning 24 h after training until the day of the retrieval test (i.e., Gαstraining group) or expressing the transgene during retrieval but not training (DOX was given 2.5 weeks before training until 24 h following training; Gαsretrieval group). In order to have sufficient time for changing the expression status of the transgene, freezing to the training context was measured 2.5 weeks following training. Gαstraining and control animals had similar levels of freezing. In contrast, compared to non-expressing control mice, freezing behavior in Gαsretrieval mice was significantly attenuated.

In addition to the behavioral tests, the study sought to analyze the effects of over-expression of the transgene on the brain morphology. MRI experiments found that Gαs expression was associated with a dramatic enlargement of the lateral ventricles, which was found to be independent of changed total brain volumes that were moderately reduced. Curiously, suppression of expression of the transgene during adulthood for as long as 6 weeks was not able to reverse the ventricular enlargement. By regulating timing of expression of the transgene, the authors found that developmental expression led to a striking enlargement of the lateral ventricles and an associated significant reduction in sizes of the dorsal and ventral striatum, and, to a lesser extent, frontal cortex and dorsal hippocampus. Similar to developmental expression, adult expression of Gαs also produced significantly enlarged lateral ventricles accompanied by a smaller dorsal and ventral striatum. Interestingly, the authors noted that the extent of the change in the lateral ventricles and ventral striatum was less prominent in Gαsadult than Gαsdev mice, suggesting the major contribution to the brain pathology of developmental expression of the transgene.

The authors also performed pharmacological and biochemical studies that suggested that the Gαs-induced behavioral deficits may be associated with compensatory decline in levels of hippocampal and cortical cyclic AMP (cAMP). These decreases in cAMP may be responsible for reduced activation of the guanine exchange factor Epac (also known as RapGEF 3/4) since the select Epac agonist, 8-pCPT-20-O-Me-cAMP, increased PPI and improved memory in mice. Thus, the authors hypothesize that the developmental impact of increased Gαs expression could lead to a specific phenotypic manifestation and that Epac could be a novel target for the treatment of both developmentally regulated and non-developmentally regulated symptoms associated with schizophrenia

Inducible expression of dopamine 2 receptor

Increased activity of D2 receptors (D2Rs) in the striatum has been linked to the pathophysiology of schizophrenia (Meisenzahl et al., 2007). To determine directly the behavioral and physiological consequences of increased D2R function in the striatum, Kellendonk et al. (2006) have generated transgenic mice with inducible expression of D2R restricted to the striatum (Kellendonk et al., 2006; Drew et al., 2007; Bach et al., 2008). I will only briefly describe this study and model. For more details, the reader is referred to Chapter 7.

Transgenic D2 receptors were functional and transgenic mice had 15% higher receptor-binding capacity than their littermates reminiscent of the 12% increase observed in schizophrenic patients (Laruelle et al., 1996). D2R transgenic mice did not exhibit reliable alterations in locomotor activity in open field, sensorimotor gating as tested by PPI, or anxiety assayed elevated plus maze. In contrast, in a series of very elegant and sophisticated cognitive tests, the authors observed moderate but significant abnormalities, consistent with cognitive impairments found in patients with schizophrenia, particularly deficits on tests of executive function, traditionally considered related to frontal lobe dysfunctions (Ross et al., 2006).

Another type of executive function abnormal in subjects with schizophrenia is attentional set-shifting and cognitive and behavioral flexibility (Kellendonk et al., 2009). An attentional set-shifting paradigm based on olfactory discrimination was used (Birrell and Brown, 2000; Colacicco et al., 2002). Although transgenic mice demonstrated normal performance in the original odor-reward association task, they were found to show significantly increased latencies to choose between the two odors during the reversal trials, indicating a subtle deficit in flexibility to reverse the original rule. D2 receptor over-expression was found to produce increased dopamine levels, decreased dopamine turnover, and increased D1 receptor activation in the medial PFC, factors which could contribute to the behavioral abnormalities.

The major advantageous feature of inducible systems is the ability to evaluate timing of the effects of expression of a transgene in question. To determine if the observed behavioral abnormalities resulted from to on-going hyperactivity of DA neurotransmission in the striatum or were rather related to developmental effects of over-expression of D2 receptors, mice were given DOX food to shut off expression in the transgene. This 2-week DOX treatment did not change the pre-existing cognitive deficit despite normalization of the binding capacity of D2 receptor in the striatum. The authors conclude that "the concurrent expression of transgenic D2 receptors in the adult animal is not responsible for the cognitive deficit; chronic or developmental expression of the receptors is sufficient to cause it" (Kellendonk et al., 2006). In order to evaluate the contribution of developmental over-expression of D2 receptors to the behavioral deficit, expression of the transgene was turned off at birth. D2 transgenic mice were found to continue to exhibit the impairment in the working memory task. Thus, the persistence of this deficit strongly suggests the critical contribution of over-expression during development.

Conclusions

Genetic mouse models are unique tools in our attempts to understand schizophrenia. To gain insights into such multi-factorial and heterogeneous neurodevelopmental disease as schizophrenia, it is critical to regulate gene expression in temporal and region- or cell-specific manners. The ability to turn gene expression on or off in the restricted cells or tissues at specific time permits unprecedented advantages in elucidating gene functions.

In this mini-review, I have discussed several inducible genetic mouse models of schizophrenia. The characterization and evaluation of these mouse models have provided valuable insights into aspects of the functions of the genes and their mutations. Mutant DISC1 is a telling example. Recent studies have clearly demonstrated that DISC1 has multiple functions (Chubb et al., 2008). As a scaffolding protein with several isoforms, DISC1 interacts with different proteins at various time points, suggesting that variable protein–protein interactions could be involved in different neurodevelopmental processes. It can explain some inconsistencies between the neuronal and behavioral data from in vitro and in vivo studies (e.g., Kamiya et al., 2005 vs. Duan et al., 2007) and emphasizes that studying timing of the effects of mutant DISC1 is of a paramount importance. In this context, inducible mutant DISC1 mouse models have shed light on variable effects of the mutation at different time points across the mouse life span (e.g., Li et al., 2007). Further use of inducible genetic mouse models will help better identify prenatal and postnatal effects of the gene, the effects of varying duration of expression, will more precisely determine which neurobehavioral features of the phenotype are reversible and which are "fixed" and "long-lasting". Another interesting implication of the multiple time-dependent functions of DISC1 is that this gene has been associated with both schizophrenia and mood disorders (Ross, 2006). One could speculate that alterations in the DISC1 functions at early stages of brain maturation may be linked to schizophrenia symptoms while changes in DISC1 expression during early or late postnatal life could produce more subtle brain pathology that is associated with affective disorders. Inducible and conditional transgenic mouse will continue to be instrumental in addressing this question as well.

The ability to regulate mutant gene expression may provide information about vulnerable time points for the generation of mental disorders. Defining critical time points may allow investigators to study the specific processes occurred during these time frames. The thorough understanding of the mechanisms underlying the vulnerable periods may lead to development of mechanism-specific therapeutics aimed to these time frames.

One also needs to discuss several important caveats of inducible genetic models. Although they are excellent systems for studying the biology of the gene, generation of these models is time-consuming and costly. Full evaluation of the effects of the transgene often requires inclusion of multiple groups of mice such as double transgenic, two types of single transgenic that are parts of the bi-transgenic model, and WT mice that do not carry any genetic modifications. If DOX is used to regulate expression of the transgene, it would be important to have a group of single transgenic mice treated with this compound also to control for possible side effects of the antibiotic. Mice that express tTA are often used as single transgenic control mice as they have been found to exhibit some behavioral abnormalities compared to WT mice of the same background (McKinney et al., 2008).

Future developments

Pioneering studies in inducible and conditional transgene expression have brought about the development of a wide variety of controlled gene expression systems. Among them, the tetracycline-controlled expression systems have been used extensively in vitro and in vivo. In recent years, some strategies derived from tetracycline-inducible system alone, as well as the combined use of Tet-based systems and Cre/lox P switching gene expression system, have been newly developed to allow for more flexibility in exploring gene functions. The reported successes in using the tetracycline-inducible system in transgenic mice have led to the development of more sophisticated strategies for manipulating target gene expression (Stieger et al., 2009).

All existing inducible genetic models have thus regulated expression of the transgenes in neurons only. However, recent studies have demonstrated that abnormalities in non-neuronal cells, e.g., astrocytes and oligodendrocytes, also play a major role in schizophrenia development (e.g., Katsel et al., 2008). Thus, future inducible and conditional mouse models of schizophrenia will likely attempt to regulate gene expression in a cell-specific manner to elucidate the gene functions in non-neuronal cells as well. Controlling gene expression in non-neuronal cells will also have an important implication for development of mouse models of schizophrenia based on combinations of genetic and environmental factors as many environmental insults are mediated by glial cells that intimately interact with neurons (Sawa et al., 2004; Ayhan et al., 2009). Elucidating these interactions across neurodevelopment is crucial for our knowledge of the complexity of schizophrenia. To date, the genetic animal models of schizophrenia mostly targeted one gene at a time. Future models may include inducible systems to regulate expression of two or more genes. Temporal and tissue-specific regulation of gene expression with RNAi technologies and regulations of expression of non-coding transcripts should also shed light on the molecular mechanisms of abnormal neurodevelopment.

Acknowledgments

The review was supported by grants from NIMH; NARSAD and Autism Speaks.

References

Arnold, S. E., Talbot, K., & Hahn, C. G. (2005). Neurodevelopment, neuroplasticity, and new genes for schizophrenia. *Progressin Brain Research*, *147*, 319–345.

Ayhan, Y., Sawa, A., Ross, C. A., & Pletnikov, M. V. (2009). Animal models of gene–environment interactions in schizophrenia. *Behavioural Brain Research*, *204*(2), 274–281.

Bach, M. E., Simpson, E. H., Kahn, L., Marshall, J. J., Kandel, E. R., & Kellendonk, C. (2008). Transient and selective overexpression of D2 receptors in the striatum causes persistent deficits in conditional associative learning. *Proceedings of the National Academy of Sciences of the United States of America*, *105*(41), 16027–16032.

Birrell, J. M., & Brown, V. J. (2000). Medial frontal cortex mediates perceptual attentional set shifting in the rat. *Journal of Neuroscience*, *20*, 4320–4324.

Brandon, N. J., Handford, E. J., Schurov, I., Rain, J. C., Pelling, M., Duran-Jimeniz, B., et al. (2004). *Molecularand Cellular Neurosciences*, *25*, 42–55.

Camargo, L. M., Collura, V., Rain, J. C., Mizuguchi, K., Hermjakob, H., Kerrien, S., et al. (2007). Disrupted in Schizophrenia 1 Interactome: Evidence for the close connectivity of risk genes and a potential synaptic basis for schizophrenia. *Molecular Psychiatry*, *12*(1), 74–86.

Chubb, J. E., Bradshaw, N. J., Soares, D. C., Porteous, D. J., & Millar, J. K. (2008). The DISC locus in psychiatric illness. *Molecular Psychiatry*, *13*(1), 36–64.

Colacicco, G., Welzl, H., Lipp, H. P., & Wurbel, H. (2002). Attentional set-shifting in mice: Modification of a rat paradigm, and evidence for strain-dependent variation. *Behavioural Brain Research*, *132*, 95–102.

Davidzon, G., Greene, P., Mancuso, M., Klos, K. J., Ahlskog, J. E., Hirano, M., et al. (2006). Early onset familial parkinsonism due to POLG mutations. *Annals of Neurology*, *59*(5), 859 62.

Degenkolb, J, Takahashi, M., Ellestad, G. A., & Hillen, W. (1991). Structural requirements of tetracycline-Tet repressor interaction: Determination of equilibrium binding constants for tetracycline analogs with the Tet repressor. *Antimicrobial Agents and Chemotherapy*, *35*, 1591–1595.

Drew, M. R., Simpson, E. H., Kellendonk, C, Herzberg, W. G., Lipatova, O., Fairhurst, S., et al. (2007). Transient overexpression of striatal D2 receptors impairs operant motivation and interval timing. *Journal of Neuroscience*, *27*(29), 7731–7739.

Duan, X., Chang, J. H., Ge, S., Faulkner, R. L., Kim, J. Y., Kitabatake, Y., et al. (2007). Disrupted-In-Schizophrenia 1 regulates integration of newly generated neurons in the adult brain. *Cell*, *130*(6), 1146–1158.

Eastwood, S. L. (2004). The synaptic pathology of schizophrenia: Is aberrant neurodevelopment and plasticity to blame. *International Reviewof Neurobiology*, *59*, 47 72.

Gingrich, J. R., & Roder, J. (1998). Inducible gene expression in the nervous system of transgenic mice. *Annual Review of Neuroscience*, *21*, 377–405.

Glantz, L. A., & Lewis, D. A. (2000). Decreased dendritic spine density on prefrontal cortical pyramidal neurons in schizophrenia. *Archivesof General Psychiatry*, *57*(1), 65–73.

Gossen, M., & Bujard, H. (1992). Tight control of gene expression in mammalian cells by tetracycline-responsive promoters. *Proceedings of the National Academy of Sciences of the United States of America*, *89*, 5547–5551.

Gossen, M, Freundlieb, S., Bender, G., Muller, G., Hillen, W., & Bujard, H. (1995). Transcriptional activation by tetracyclines in mammalian cells. *Science*, *268*, 1766–1769.

Hikida, T, Jaaro-Peled, H, Seshadri, S, Oishi, K, Hookway, C, Kong, S, et al. (2007). Dominant-negative DISC1 transgenic mice display schizophrenia-associated phenotypes detected by measures translatable to humans. *Proceedings of the National Academy of Sciences of the United States of America*, *104*(36), 14501–14506.

Hillen, W, & Berens, C. (1994). Mechanisms underlying expression of Tn10 encoded tetracycline resistance. *Annual Review of Microbiology*, *48*, 345–369.

Hillen, W., Klock, G., Kaffenberger, I., Wray, L. V., & Reznikoff, W. S. (1982). Purification of the TET repressor and TET operator from the transposon Tn10 and characterization of their interaction. *Journal of Biological Chemistry*, *257*, 6605–6613.

Ishizuka, K., Paek, M., Kamiya, A., & Sawa, A. (2006). A review of Disrupted-In-Schizophrenia-1 (DISC1): Neurodevelopment, cognition, and mental conditions. *Biological Psychiatry*, *59*(12), 1189–1197.

Kamiya, A., Kubo, K., Tomoda, T., Takaki, M., Youn, R., Ozeki, Y., et al. (2005). A schizophrenia-associated mutation of DISC1 perturbs cerebral cortex development. *Nature Cell Biology*, *12*, 1167–1178.

Kamiya, A., Tomoda, T., Chang, J., Takaki, M., Zhan, C., Morita, M., et al. (2006). DISC1-NDEL1/NUDEL protein interaction, an essential component for neurite outgrowth, is modulated by genetic variations of DISC1. *Human Molecular Genetics*, *15*, 3313–3323.

Katsel, P., Davis, K. L., Li, C., Tan, W., Greenstein, E., Kleiner Hoffman, L. B., et al. (2008). Abnormal indices of cell cycle activity in schizophrenia and their potential association with oligodendrocytes. *Neuropsychopharmacology*, *33*(12), 2993–3009.

Kellendonk, C., Simpson, E. H., & Kandel, E. R. (2009). Modeling cognitive endophenotypes of schizophrenia in mice. *Trends in Neurosciences*, *32*(6), 347–358.

Kellendonk, C., Simpson, E. H., Polan, H. J., Malleret, G., Vronskaya, S., Winiger, V., et al. (2006). Transient and selective overexpression of dopamine D2 receptors in the striatum causes persistent abnormalities in prefrontal cortex functioning. *Neuron*, *49*(4), 603–615.

Kelly, M. P., Stein, J. M., Vecsey, C. G., Favilla, C., Yang, X., Bizily, S. F., et al. (2009). Developmental etiology for neuroanatomical and cognitive deficits in mice overexpressing Galphas, a G-protein subunit genetically linked to schizophrenia. *Molecular Psychiatry*, (4), 398–415.

Kessler, R. C., Amminger, G. P., Aguilar-Gaxiola, S., Alonso, J., Lee, S., & Ustün, T. B. (2007). Age of onset of mental disorders: A review of recent literature. *Current Opinion in Psychiatry*, *20*(4), 359 64.

Kessler, R. C., Birnbaum, H., Demler, O., Falloon, I. R., Gagnon, E., Guyer, M., et al. (2005). The prevalence and correlates of nonaffective psychosis in the National Comorbidity Survey Replication (NCS-R). *Biological Psychiatry*, *58* (8), 668–676.

Kida, S., Josselyn, S. A., de Ortiz, S. P., Kogan, J. H., Chevere, I., Masushige, S., et al. (2002). CREB required for the stability of new and reactivated fear memories. *Nature Neuroscience*, *5*, 348–355.

Kistner, A., Gossen, M., Zimmermann, F., Jerecic, J., Ullmer, C., et al. (1996). Doxycycline-mediated quantitative and tissue-specific control of gene expression in transgenic mice. *Proceedings of the National Academy of Sciences of the United States of America*, *93*, 10933–10938.

Laruelle, M., Abi-Dargham, A., van Dyck, C. H., Gil, R., D'Souza, C. D., Erdos, J., et al. (1996). Single photon emission computerized tomography imaging of amphetamine-induced dopamine release in drug-free schizophrenic subjects. *Proceedings of the National Academy of Sciences of the United States of America, 93*, 9235–9240.

Lewis, D. A., & Levitt, P. (2002). Schizophrenia as a disorder of neurodevelopment. *Annual Review of Neuroscience, 25*, 409–432.

Li, W., Zhou, Y., Jentsch, J. D., Brown, R. A., Tian, X., Ehninger, D., et al. (2007). Specific developmental disruption of disrupted-in-schizophrenia-1 function results in schizophrenia-related phenotypes in mice. *Proceedings of the National Academy of Sciences of the United States of America, 104*(46), 18280–18285.

Lieberman, J. A., Stroup, T. S., McEvoy, J. P., Swartz, M. S., Rosenheck, R. A., Perkins, D. O., Davis, R. S., Davis, C. E., Lebowitz, B. D., Severe, J., & Hsiao, J. K. (2005). Effectiveness of antipsychotic drugs in patients with chronic schizophrenia. *N Engl J Med, 353*, 1209–1223.

Logie, C., & Stewart, F. (1995). Ligand-regulated site-specific recombination. *Proceedings of the National Academy of Sciences of the United States of America, 92*, 5940–5944.

Mallo, M. (2006). Controlled gene activation and inactivation in the mouse. *Frontiers in Bioscience, 11*, 313–327.

Marenco, S., & Weinberger, D. R. (2000). The neurodevelopmental hypothesis of schizophrenia: Following a trail of evidence from cradle to grave. *Development and Psychopathology, 12*(3), 501–527.

McKinney, B. C., Schneider, J. S., Schafer, G. L., Lowing, J. L., Mohan, S., Zhao, M. X., et al. (2008). Decreased locomotor activity in mice expressing tTA under control of the CaMKII alpha promoter. *Genes Brain and Behaviour, 7*(2), 203–213.

Meisenzahl, E. M., Schmitt, G. J., Scheuerecker, J., & Möller, H. J. (2007). The role of dopamine for the pathophysiology of schizophrenia. *International Reviewof Psychiatry, 19*(4), 337–345.

Metzger, D., Clifford, J., Chiba, H., & Chambon, P. (1995). Conditional site-specific recombination in mammalian cells using a ligand-dependent chimeric Cre recombinase. *Proceedings of the National Academy of Sciences of the United States of America, 92*, 6991–6995.

Millar, J. K., Christie, S., Anderson, S., Lawson, D., Hsiao-Wei Loh, D., Devon, R. S., et al. (2001). Genomic structure and localization within a linkage hotspot of Disrupted In Schizophrenia 1, a gene disrupted by a translocation segregating with schizophrenia. *Molecular Psychiatry, 6*(2), 173–178.

Mohn, A. R., Gainetdinov, R. R., Caron, M. G., & Koller, B. H. (1999). Mice with reduced NMDA receptor expression display behaviors related to schizophrenia. *Cell, 98*(4), 427–436.

Morris, J. A., Kandpal, G., Ma, L., & Austin, C. P. (2003). DISC1 (Disrupted-In-Schizophrenia 1) is a centrosome-associated protein that interacts with MAP1A, MIPT3, ATF4/5 and NUDEL: Regulation and loss of interaction with mutation. *Human Molecular Genetics, 12*, 1591–1608.

Moy, S. S., Nadler, J. J., Perez, A., Barbaro, R. P., Johns, J. M., Magnuson, T. R., et al. (2004). *Genes Brain and Behaviour, 3*, 287–302.

Ozeki, Y., Tomoda, T., Kleiderlein, J., Kamiya, A., Bord, L., Fujii, K., et al. (2003). *Proceedings of the National Academy of Sciences of the United States of America, 100*, 289–294.

Pagsberg, A. K., Baaré, W. F., Raabjerg Christensen, A. M., Fagerlund, B., Hansen, M. B., Labianca, J., et al. (2007). Structural brain abnormalities in early onset first-episode psychosis. *Journal of Neural Transmission, 114*(4), 489–498.

Piscopo, P., Marcon, G., Piras, M. R., Crestini, A., Campeggi, L. M., Deiana, E., et al. (2008). A novel PSEN2 mutation associated with a peculiar phenotype. *Neurology, 70*(17), 1549–1554.

Pletnikov, M. V., Ayhan, Y., Nikolskaia, O., Xu, Y., Ovanesov, M. V., Huang, H., et al. (2008). Inducible expression of mutant human DISC1 in mice is associated with brain and behavioral abnormalities reminiscent of schizophrenia. *Molecular Psychiatry, 13*(2), 173–186.

Ross, C. A., Margolis, R. L., Reading, S. A., Pletnikov, M., & Coyle, J. T. (2006). Neurobiology of schizophrenia. *Neuron, 52*(1), 139–153.

Sawa, A., Pletnikov, M. V., & Kamiya, A. (2004). Neuron–glia interactions clarify genetic environmental links in mental illness. *Trends in Neurosciences, 27*(6), 294–297.

Shen, S., Lang, B., Nakamoto, C., Zhang, F., Pu, J., Kuan, S. L., et al. (2008). Schizophrenia-related neural and behavioral phenotypes in transgenic mice expressing truncated Disc1. *Journalof Neuroscience, 28*(43), 10893–10904.

Stieger, K., Belbellaa, B., Le Guiner, C., Moullier, P., & Rolling, F. (2009). In vivo gene regulation using tetracycline-regulatable systems. *Advanced Drug Delivery Reviews, 61*, 527–541.

Stone, M. H. (2006). History of schizophrenia and its antecedents. In JA Lieberman, S Stroup, & DO. Perkins (Eds.), *Textbook of Psychiatry*. American Psychiatric Publishing.

Straub, R. E., & Weinberger, D. R. (2006). Schizophrenia genes-famine to feast. *Biological Psychiatry, 60*(2), 81–83.

Sullivan, P. F., Kendler, K. S., & Neale, M. C. (2003). Schizophrenia as a complex trait: Evidence from meta-analysis of twin studies. *Archivesof General Psychiatry, 60* (12), 1187–1192.

Vita, A., De Peri, L., Silenzi, C., & Dieci, M. (2006). Brain morphology in first-episode schizophrenia: A meta-analysis of quantitative magnetic resonance imaging studies. *Schizophrenia Research, 82*(1), 75–88.

Weinberger, D. R. (1996). On the plausibility of "the neurodevelopmental hypothesis" of schizophrenia. *Neuropsychopharmacology, 14*(3 Suppl), 1S–11S.

Woods, B. T. (1998). Is schizophrenia a progressive neurodevelopmental disorder? Toward a unitary pathogenetic mechanism. *American Journalof Psychiatry, 155*(12), 1661–1670.

Wu, E. Q., Birnbaum, H. G., Shi, L., Ball, D. E., Kessler, R. C., Moulis, M., & Aggarwal, J. (2005). The economic burden of schizophrenia in the United States in 2002. *J Clin Psychiatry, 66*, 1122–1129.

Wu, E. Q., Shi, L., Birnbaum, H., Hudson, T., & Kessler, R. (2006). Annual prevalence of diagnosed schizophrenia in the

USA: A claims data analysis approach. *Psychological Medicine*, *36*(11), 1535–1540.

Yamamoto, A., Lucas, J. J., & Hen, R. (2000). Reversal of neuropathology and motor dysfunction in a conditional model of Huntington's disease. *Cell*, *101*(1), 57–66.

Zhang, Y., Riesterer, C., Ayrall, A.-M., Sablitzky, F., Littlewood, T. D., & Reth, M. (1996). Inducible site directed recombination in mouse embryonic stem cells. *Nucleic Acids Research*, *24*, 543–548.

Zhu, Z., Zheng, T., Lee, C. G., Homer, R. J., & Elias, J. A. (2002). Tetracycline-controlled transcriptional regulation systems: Advances and application in transgenic animal modeling. *Seminarsin Cell & Developmental Biology*, *13*(2), 121–128.

SECTION III

Representative Models

A. Sawa (Ed.)
Progress in Brain Research, Vol. 179
ISSN 0079-6123

CHAPTER 6

NR1 knockdown mice as a representative model of the glutamate hypothesis of schizophrenia

Amy J. Ramsey*

University of Toronto, Department of Pharmacology and Toxicology, Toronto, ON, Canada

Abstract: *N*-methyl d-aspartate (NMDA) receptor subunit NR1 knockdown (NR1-KD) mice have a global reduction of NMDA receptors, enabling their use as a genetic model to study the role of NMDA receptors in the pathophysiology of schizophrenia. This targeted mutation results in a spectrum of altered behaviors that are similar to those induced by NMDA receptor antagonists, which have long been used to model schizophrenia in animals. NR1-KD mice serve as a complementary tool to pharmacological models, providing insight into the consequences of sustained NMDA receptor dysfunction in early brain development and throughout the life of the animal. Though in many respects the phenotype of NR1-KD mice mimics that of acute NMDA receptor antagonism, there are also notable differences. In this chapter we highlight some of the molecular, behavioral, and neurophysiological phenotypes of NR1-KD mice and compare these to pharmacological models of NMDA receptor dysfunction. Through the study of these models, our improved understanding of how the brain adapts to persistent NMDA receptor hypofunction may eventually suggest new therapeutic strategies for schizophrenia.

Keywords: NMDA receptor; glutamate; behavioral pharmacology; pharmacology; physiology

The etiology of schizophrenia defies simple explanation and remains unclear. It can be described as a genetic disease with a non-mendelian inheritance pattern that is influenced by environmental factors. Schizophrenia may result from a single, rare mutation of strong effect (Millar et al., 2000; Walsh et al., 2008), or from the inherited combination of an unknown number of susceptibility alleles with individually modest effect (International Schizophrenia Consortium et al., 2009). The genes that have been identified to date suggest that no one neurotransmitter system can explain all cases of schizophrenia, although these gene products may functionally converge in the neurocircuitry of the brain (Harrison and Weinberger, 2005; Lisman et al., 2008). Appreciating the spectrum of genetic and environmental events that can result in schizophrenia means generating many genetic animal models to understand the biology of this disease.

In this chapter we discuss the application of *N*-methyl d-aspartate (NMDA) receptor deficient mice as a model to study schizophrenia. The NR1 knockdown (NR1-KD) mouse does not model a naturally occurring schizophrenia susceptibility gene. Instead, it is used to model a state of NMDA receptor hypofunction. Several lines of evidence implicate NMDA receptors either in the cause of schizophrenia or in the pathophysiological

*Corresponding author.
Tel.: +1 416 978 2509; Fax: +1 416 978 6395;
E-mail: a.ramsey@utoronto.ca

DOI: 10.1016/S0079-6123(09)17906-2

manifestations of the disease. As enumerated in several excellent reviews, including Coyle, 2006 and Javitt, 2007, these lines of evidence include genetic studies (Harrison and Weinberger, 2005), imaging studies (Pilowsky et al., 2006), postmortem analysis of schizophrenic brains (Gao et al., 2000; Kristiansen et al., 2006), and, most compellingly, the pharmacological properties of NMDA receptor antagonists and co-agonists in schizophrenics and in healthy subjects (Coyle et al., 2002). In fact, NMDA receptor antagonists can induce in normal humans behaviors that resemble both psychotic (positive) and deficit (negative) symptoms of schizophrenia (Luby et al., 1959; Krystal et al., 1994; Lahti et al., 2001). Furthermore, these drugs can exacerbate or precipitate symptoms in schizophrenic patients (Lahti et al., 1995). These findings have been the cornerstone of the glutamate hypothesis of schizophrenia.

Originating from the pharmacology of NMDA receptor antagonists in humans, there is a large body of literature describing the use of phencyclidine (PCP), ketamine, and dizocilpine (MK-801) in animals to model schizophrenia. These models have primarily been used to identify drugs with antipsychotic properties, based on their ability to reverse PCP- or MK-801-induced behaviors. However, they have also been used to test the validity of the NMDA receptor hypofunction hypothesis of schizophrenia to determine whether PCP or MK-801 administration can induce changes in brain function that are consistent with the changes observed in schizophrenia (Jentsch and Roth, 1999).

A genetic model of NMDA receptor hypofunction

To complement and extend these pharmacological studies, NR1-KD mice were among the first to be described as representative for a genetic model of NMDA receptor hypofunction (Mohn et al., 1999). Null mutation of *Grin1*, the gene that encodes the essential NR1 subunit of NMDA receptors, is lethal (Forrest et al., 1994; Li et al., 1994). However, NR1-KD mice are viable because they have only a partial loss-of-function, or hypomorphic mutation, of *Grin1*. The hypomorphic mutation is achieved by the targeted insertion of nearly 2 kb of foreign DNA into an intron of *Grin1*, near the 3′ end of the gene (Fig. 1). The foreign DNA is a neomycin selectable marker, and its presence reduces the amount of full-length mRNA that is produced; this is likely due to premature termination at the polyadenylation sequence of *neo*. In the remaining full-length transcript that is produced, the *neo* sequence is removed by splicing to generate functional subunits in significantly reduced quantity. Because assembly of NMDA receptors requires both NR1 and NR2 subunits, the deficiency in NR1 protein leads to a reduction in the amount of NR2 subunit (Ramsey et al., 2008), and in the amount of heteromeric receptor (Mohn et al., 1999).

There are consequently sufficient levels of NR1 subunit to support survival, but the levels of NMDA receptor are estimated to be only 10–20% of normal levels as measured by radioligand binding with [^{3}H] MK-801 (Mohn et al., 1999; Duncan et al., 2002). In wild-type brain there are regional differences in the levels of NR1 and NMDA receptors, with the highest levels being in the hippocampus and forebrain (Monaghan et al., 1989). However, in the NR1-KD brain, the mutation uniformly limits the quantity of receptor that can be produced; the reduction in receptors is consequently most significant in those brain regions where NMDA receptors are normally most abundant. In hippocampus there is a 90% reduction in NMDA receptors, but in brainstem there is only a 60% reduction, which may explain why mutant mice are able to suckle, to breathe, and hence to avoid the perinatal lethality of null mutation. The effect of the mutation on NMDA receptor levels in specific brain regions is described further in Fig. 1 and Table 1.

Comparing genetic and pharmacological models of NMDA receptor hypofunction

It is important to keep the region-selective deficits of NMDA receptors in mind when comparing NR1-KD mice to wild-type mice that have been treated with non-competitive NMDA receptor antagonists. The physiological effects of PCP and MK-801 are dose-dependent; low doses have

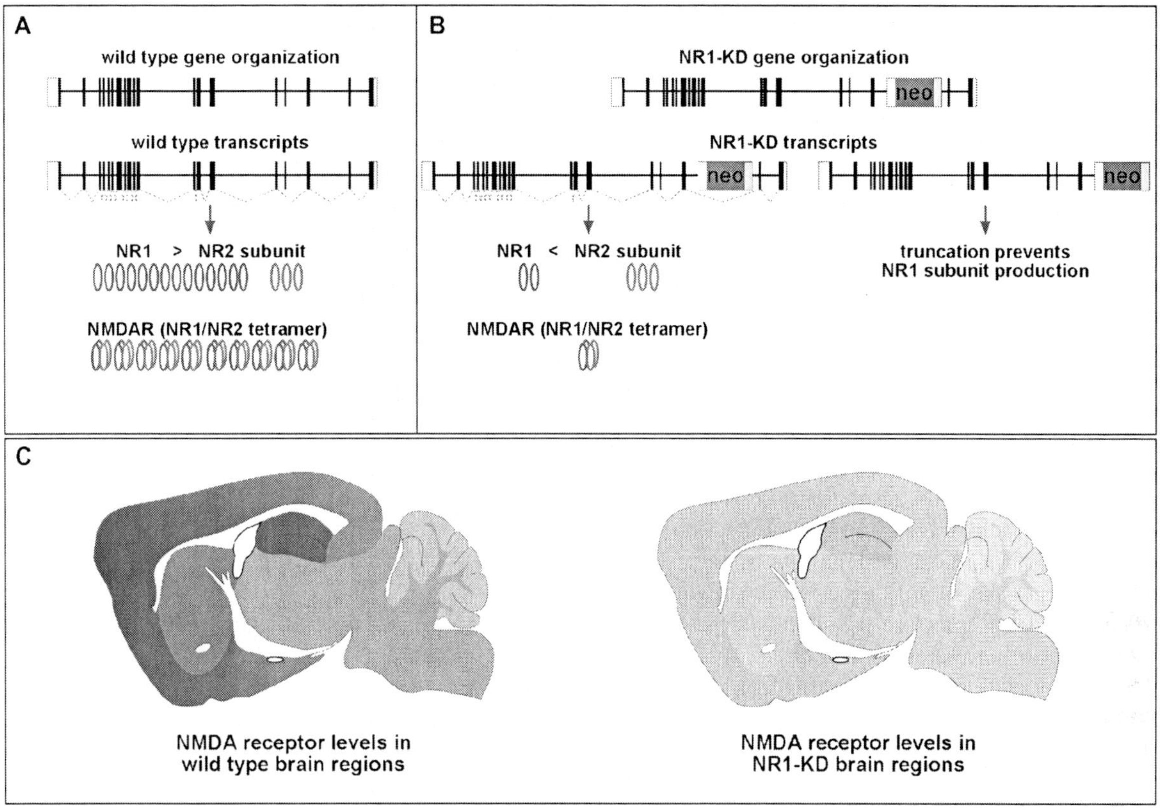

Fig. 1. Hypomorphic *mutation of Grin1 reduces NMDA receptor subunit NR1 message and protein.* In wild-type mice (A), *Grin1* is transcribed and alternately spliced. NR1 subunits are produced in excess of NR2 subunits, which limit the amount of functional receptor that can be produced (Huh and Wenthold, 1999). In NR1-KD mice (B), the presence of *neo* leads to the production of transcripts that are truncated due to the polyadenylation signal of the *neo* gene. For those transcripts that are full length, proper splicing of the message removes *neo* and produces functional NR1 subunits. NR1 levels are reduced and are no longer in excess of NR2 subunits, producing fewer functional NMDA receptors. Although mutation of NR1 leads to global reductions in NMDA receptors, the deficit is most apparent in brain regions where NMDA receptors are normally highest (C). This illustration depicts the [^{3}H] MK-801 radioligand binding results, where darker shades represent higher levels of receptor.

Table 1. Regional impact of NR1-KD hypomorphic mutation on NMDA receptor levels

Brain region	WT MK-801 Bmax	NR1-KD MK-801 Bmax	Regional deficit
Hippocampus	1970 ± 62 fmol/mg	207 ± 73 fmol/mg	10% of WT
Frontal cortex	1320 ± 82 fmol/mg	187 ± 54 fmol/mg	14% of WT
Striatum	780 ± 11 fmol/mg	165 ± 16 fmol/mg	21% of WT
Brainstem	120 ± 15 fmol/mg	48 ± 10 fmol/mg	40% of WT

NMDA receptor levels are measured by competition binding with cold and tritiated MK-801 ($n = 3$ animals for each genotype) to determine the maximal binding (Bmax) as described in Mohn et al. (1999).

psychotomimetic effects, moderate doses induce ataxia and dissociative anesthesia, high doses suppress breathing and can cause death (Javitt and Zukin, 1991). In mice, low doses of PCP (3–10 mg/kg) or MK-801 (0.2–0.3 mg/kg) are used to model psychosis.

These doses induce a number of altered behaviors reflecting decreased NMDA receptor transmission. Those that have been best characterized and routinely used in evaluating new antipsychotics include increased locomotor activity, decreased habituation to a novel environment, increased stereotypic (repetitive) behaviors, decreased performance in spatial and working memory tasks, decreased sensorimotor gating, and reduced sociability.

In these behaviors, NR1-KD mice display abnormal phenotypes that are qualitatively and quantifiably similar to MK-801-treated mice (Mohn et al., 1999; Duncan et al. 2004; Dzirasa et al., 2009; Halene et al., 2009). In addition, NR1-KD mice have an exaggerated startle response to auditory stimulus (Duncan et al., 2006). In sociability, stereotypy, and habituation to novel environment, the quantitative measures of these behaviors are remarkably similar between the genetic and pharmacological models. In contrast, NR1-KD mice do not display ataxia (Mohn et al., 1999), a behavior that reflects impaired function of the cerebellum, and is induced by higher doses of MK-801 (0.5 mg/kg). Thus the behavioral profile of NR1-KD mice most resembles that of mice treated with low doses of MK-801 or PCP (psychotomimetic), rather than higher doses that induce ataxia. The reason for this may be the regional impact of the mutation on brain regions that have higher levels of NMDA receptors and would normally be more affected by low doses of NMDA receptor antagonists.

Behavioral pharmacology of NR1-KD mice

The characterization of NR1-KD behavioral pharmacology provides information on the status of neurotransmitter systems, and also lends support for the utility of the model in the discovery of novel antipsychotics. As with MK-801-treated mice, some of the behavioral abnormalities of NR1-KD mice can be ameliorated to varying degrees with antipsychotics or psychoactive drugs. Locomotor activity in NR1-KD mice is higher than wild-type mice; this is due to a failure to habituate to novel environment (Mohn et al., 1999; Halene et al., 2009). This behavioral readout is most easily amenable to evaluating drugs with antipsychotic action, and also has construct validity because exploratory ambulation is regulated by neurotransmitters and neurocircuits that have been implicated in schizophrenia. NR1-KD hyperactivity is attenuated by first- and second-generation antipsychotics haloperidol, clozapine, olanzapine, and by the D2 dopamine receptor agonist quinpirole (Mohn et al., 1999; Duncan et al., 2006; Ramsey et al., 2008). The same antipsychotic doses that attenuate MK-801-induced hyperactivity are also effective in NR1-KD mice. It is likely that, in the behavioral paradigm of locomotor activity in a novel environment, the majority of drugs that reverse MK-801-induced hyperactivity will also attenuate NR1-KD hyperactivity.

It is of interest to determine whether non-dopaminergic drugs will be effective in normalizing the exploratory behavior of these mice. Because their hyperactivity is a result of delayed habituation, it may reflect a deficit in cognition (Nilsson et al., 2001) rather than an overactivation of dopamine neurons. Although further studies are warranted, the initial characterization of dopamine transmission in NR1-KD mice is not suggestive of a hyperdopaminergic state. In fact dopamine D1 and D2 receptor numbers, and behavioral responses to amphetamine and D1 or D2 agonists are unchanged, as are the basal and amphetamine-stimulated levels of extracellular dopamine (Mohn et al., 1999; Ramsey et al., 2008). The neurochemical response to amphetamine is notable, given the finding that schizophrenics have a greater amphetamine-induced dopamine efflux than normal volunteers (Laruelle et al., 1996). This aspect of the NR1-KD phenotype highlights a limitation of the model, but also suggests that the model is more suitable for discovery of drugs that improve cognition and negative symptoms rather than those that manage dopamine-mediated psychosis.

Indeed the NR1-KD phenotype of reduced sociability, or social withdrawal, is robust and consistently observed in different laboratories (Mohn et al., 1999; Duncan et al., 2004, 2009; Halene et al., 2009). With the development of a rapid screen for sociability (Moy et al., 2004), this behavior, like exploratory ambulation, is amenable to evaluate drugs that improve cognition or sociability. Clozapine improves NR1-KD social avoidance behaviors (Mohn et al., 1999) and further studies with other drugs are warranted.

Neurophysiology of NR1-KD mice: sensory processing and gamma oscillations

Neurophysiological measurements of brain function are measured in mice using implanted electrodes, and provide one of the few endophenotypes that can be directly translated from humans to mice (Javitt et al., 2008; Bickel and Javitt, 2009). Electrical recording of brain activity provides information on how local neuron groups respond to sensory stimuli, and how neurons in the same brain structure or across different brain structures synchronize activity. Sensory information processing is also measured in the context of sensorimotor gating through the paradigm of prepulse inhibition of acoustic startle response (PPI).

Deficient sensory processing has been reported in NR1-KD mice in the PPI paradigm (Duncan et al., 2004; Fradley et al., 2005; Bickel et al., 2008). Analysis of event-related potentials evoked by auditory and visual stimuli has also been performed (Bickel et al., 2007, 2008; Halene et al., 2009). These studies report significant changes in the NR1-KD patterns of electrical activity in response to auditory or visual stimuli; however, these patterns are different from what is observed after NMDA receptor antagonist treatment. The explanation for this discrepancy may lie in the chronic NMDA receptor deficiency throughout development in NR1-KD mice. It may also be the case that pharmacological blockade of NMDA receptors preferentially affects a subset of neuronal circuits than the more global effect of the genetic mutation.

While these studies have implications for the temporal processing of sensory information, a more recent study has implications for the mechanisms by which the brain binds neuronal activity from different structures together during cognitive tasks. Dzirasa et al. (2009) demonstrated changes in NR1-KD mice in the coherence or phase coupling of gamma and theta oscillations. The rhythmic firing of neuron ensembles is described as gamma or theta oscillations based on their frequency (for example 20–80 Hz), and the interstructural synchronization of these oscillations is enhanced during cognitive tasks (Siapas et al., 2005; Sirota et al., 2008). In the study by Dzirasa et al., multiple recording electrodes were placed in different structures of the brain, including prefrontal cortex, hippocampus, and prelimbic cortex, and oscillations are simultaneously detected at different frequencies and in different brain regions. Coupling of gamma oscillations to other firing rhythms, like theta oscillations, was measured by recording local field potentials while mice explored novel environments or remained in their home cage. The authors report that NR1-KD show a profound deficit in the interstructural (hippocampal–cortical) phase coupling between theta and gamma oscillations. Furthermore, exploration of a novel environment does not enhance synchrony as it does in wild-type mice. These findings may have relevance for schizophrenia, where disruptions in neural phase synchrony have been reported (Winterer et al., 2000; Yeragani et al., 2006).

NR1-KD as a representative model of NMDA receptor hypofunction

In the past decade the number of genetic mouse models of NMDA receptor hypofunction have steadily increased, which should not be surprising considering the many ways that NMDA receptors can be regulated. A recent review highlights the several examples of mice with targeted mutations in the genes encoding glutamate receptors (Inta et al., 2009); many of these have overlapping phenotypes with the NR1-KD mice or with each other that underscore the common consequences of NMDA receptor dysfunction or alterations in glutamate transmission. Beyond the glutamate receptor genes are the multitude of genes that regulate glutamate and glycine levels, receptor trafficking and post-translational regulation, and calcium signal transduction. Although by no means exhaustive, a short list of models for these regulatory genes includes those that have been implicated in schizophrenia by genetic or postmortem studies. Mutants for serine racemase (Basu et al., 2009; Labrie et al., 2009), glutamate carboxypeptidase II (Han et al., 2009), calcineurin (Miyakawa et al., 2003), and neuregulin 1 (Stefansson et al., 2002, and see accompanying chapter of this issue) all have indications of impaired NMDA receptor function, and display altered phenotypes relevant to schizophrenia. In this sense the NR1-KD mouse is merely a representative in the much larger effort to understand the physiological consequences of sustained NMDA receptor hypofunction that can be achieved with genetic models.

Acknowledgments

I would like to thank Ali Salahpour for critical reading and helpful discussion, and Marc Caron, Raul Gainetdinov, Michael Didriksen, Akira Sawa, and Beverly Koller for discussions of the NR1-KD mice as a model for schizophrenia. This work was supported by a NARSAD Young Investigator Award to AJR.

References

Basu, A. C., Tsai, G. E., Ma, C. L., Ehmsen, J. T., Mustafa, A. K., Han, L., et al. (2009). Targeted disruption of serine racemase affects glutamatergic neurotransmission and behavior. *Molecular Psychiatry*, *14*(7), 719–727.

Bickel, S., & Javitt, D. C. (2009). Neurophysiological and neurochemical animal models of schizophrenia: Focus on glutamate. *Behavioral Brain Research*, *204*(2), 352–362.

Bickel, S., Lipp, H. P., & Umbricht, D. (2007). Impaired attentional modulation of auditory evoked potentials in N-methyl-d-aspartate NR1 hypomorphic mice. *Genes, Brain and Behavior*, *6*(6), 558–568.

Bickel, S., Lipp, H. P., & Umbricht, D. (2008). Early auditory sensory processing deficits in mouse mutants with reduced NMDA receptor function.. *Neuropsychopharmacology*, *33* (7), 1680–1689.

Coyle, J. T. (2006). Substance use disorders and schizophrenia: A question of shared glutamatergic mechanisms. *Neurotoxicity Research*, *10*(3), 221–233.

Coyle, J. T., Tsai, G., & Goff, D. C. (2002). Ionotropic glutamate receptors as therapeutic targets in schizophrenia.. *Current Drug Targets: CNS and Neurological Disorders*, *1* (2), 183–189.

Duncan, G. E., Inada, K., Farrington, J. S., Koller, B. H., & Moy, S. S. (2009). Neural activation deficits in a mouse genetic model of NMDA receptor hypofunction in tests of social aggression and swim stress. *Brain Research*, *1265*, 186–195.

Duncan, G. E., Moy, S. S., Lieberman, J. A., & Koller, B. H. (2006). Typical and atypical antipsychotic drug effects on locomotor hyperactivity and deficits in sensorimotor gating in a genetic model of NMDA receptor hypofunction. *Pharmacology Biochemistry and Behavior*, *85*(3), 481–491.

Duncan, G. E., Moy, S. S., Perez, A., Eddy, D. M., Zinzow, W. M., Lieberman, J. A., et al. (2004). Deficits in sensorimotor gating and tests of social behavior in a genetic model of reduced NMDA receptor function. *Behavioral Brain Research*, *153*(2), 507–519.

Duncan, G., Miyamoto, S., Gu, H., Lieberman, J., Koller, B., & Snouwaert, J. (2002). Alterations in regional brain metabolism in genetic and pharmacological models of reduced NMDA receptor function. *Brain Research*, *951*(2), 166–176.

Dzirasa, K., Ramsey, A. J., Takahashi, D. Y., Stapleton, J., Potes, J. M., Williams, J. K., et al. (2009). Hyperdopaminergia and NMDA receptor hypofunction disrupt neural phase signaling. *Journal of Neuroscience*, *29*(25), 8215–8224.

Forrest, D., Yuzaki, M., Soares, H. D., Ng, L., Luk, D. C., Sheng, M., et al. (1994). Targeted disruption of NMDA receptor 1 gene abolishes NMDA response and results in neonatal death.. *Neuron*, *13*(2), 325–338.

Fradley, R. L., O'Meara, G. F., Newman, R. J., Andrieux, A., Job, D., & Reynolds, D. S. (2005). STOP knockout and NMDA NR1 hypomorphic mice exhibit deficits in sensorimotor gating. *Behavioral Brain Research*, *163*(2), 257–264.

Gao, X. M., Sakai, K., Roberts, R. C., Conley, R. R., Dean, B., & Tamminga, C. A. (2000). Ionotropic glutamate receptors and expression of *N*-methyl-d-aspartate receptor subunits in subregions of human hippocampus: Effects of schizophrenia. *American Journal of Psychiatry*, *157*(7), 1141–1149.

Halene T. B., Ehrlichman R. S., Liang Y., Christian E. P., Jonak G. J., Gur T. L., et al. (2009). Assessment of NMDA receptor NR1 subunit hypofunction in mice as a model for schizophrenia. *Genes, Brain and Behavior*, Epub ahead of print.

Han, L., Picker, J. D., Schaevitz, L. R., Tsai, G., Feng, J., Jiang, Z., et al. (2009). Phenotypic characterization of mice heterozygous for a null mutation of glutamate carboxypeptidase II.. *Synapse*, *63*(8), 625–635.

Harrison, P. J., & Weinberger, D. R. (2005). Schizophrenia genes, gene expression, and neuropathology: On the matter of their convergence. *Molecular Psychiatry*, *10*(1), 40–68.

Huh, K. H., & Wenthold, R. J. (1999). Turnover analysis of glutamate receptors identifies a rapidly degraded pool of the *N*-methyl-d-aspartate receptor subunit, NR1, in cultured cerebellar granule cells. *Journal of Biological Chemistry*, *274* (1), 151–157.

Inta D., Monyer H., Sprengel R., Meyer-Lindenberg A., Gass P. (2009). Mice with genetically altered glutamate receptors as models of schizophrenia: A comprehensive review. *Neuroscience and Biobehavioral Reviews*, August 3, Epub ahead of print.

International Schizophrenia Consortium, , Purcell, S. M., Wray, N. R., Stone, J. L., Visscher, P. M., O'Donovan, M. C., et al. (. Common polygenic variation contributes to risk of schizophrenia and bipolar disorder. *Nature*, *460*(7256), 748–752.

Javitt, D. C. (2007). Glutamate and schizophrenia: Phencyclidine, *N*-methyl-d-aspartate receptors, and dopamine–glutamate interactions. *International Review of Neurobiology*, *78*, 69–108.

Javitt, D. C., Spencer, K. M., Thaker, G. K., Winterer, G, & Hajós, M. (2008). Neurophysiological biomarkers for drug development in schizophrenia. *Nature Reviews Drug Discovery*, *7*(1), 68–83.

Javitt, D. C., & Zukin, S. R. (1991). Recent advances in the phencyclidine model of schizophrenia. *American Journal of Psychiatry*, *148*(10), 1301–1308.

Jentsch, J. D., & Roth, R. H. (1999). The neuropsychopharmacology of phencyclidine: From NMDA receptor hypofunction to the dopamine hypothesis of schizophrenia. *Neuropsychopharmacology*, *20*(3), 201–225.

Kristiansen, L. V., Beneyto, M., Haroutunian, V., & Meador-Woodruff, J. H. (2006). Changes in NMDA receptor subunits and interacting PSD proteins in dorsolateral prefrontal and anterior cingulate cortex indicate abnormal regional expression in schizophrenia. *Molecular Psychiatry*, *11*, 737–747.

Krystal, J. H., Karper, L. P., Seibyl, J. P., Freeman, G. K., Delaney, R., Bremner, J. D., et al. (1994). Subanesthetic effects of the noncompetitive NMDA antagonist, ketamine, in humans. Psychotomimetic, perceptual, cognitive, and neuroendocrine responses. *Archives of General Psychiatry*, *51*(3), 199–214.

Labrie, V., Fukumura, R., Rastogi, A., Fick, L. J., Wang, W., Boutros, P. C., et al. (2009). Serine racemase is associated with schizophrenia susceptibility in humans and in a mouse model. *Human Molecular Genetics*, *18* (17), 3227–3243.

Lahti, A. C., Koffel, B., LaPorte, D., & Tamminga, C. A. (1995). Subanesthetic doses of ketamine stimulate psychosis in schizophrenia.. *Neuropsychopharmacology*, *13*(1), 9–19.

Lahti, A. C., Weiler, M. A., Tamara Michaelidis, B. A., Parwani, A., & Tamminga, C. A. (2001). Effects of ketamine in normal and schizophrenic volunteers.. *Neuropsychopharmacology*, *25*(4), 455–467.

Laruelle, M., Abi-Dargham, A., van Dyck, C. H., Gil, R., D'Souza, C. D., Erdos, J., et al. (1996). Single photon emission computerized tomography imaging of amphetamine-induced dopamine release in drug-free schizophrenic subjects. *Proceedings of the National Academy of Sciences of the United States of America*, *93*(17), 9235–9240.

Li, Y., Erzurumlu, R. S., Chen, C., Jhaveri, S., & Tonegawa, S. (1994). Whisker-related neuronal patterns fail to develop in the trigeminal brainstem nuclei of NMDAR1 knockout mice. *Cell*, *76*(3), 427–437.

Lisman, J. E., Coyle, J. T., Green, R. W., Javitt, D. C., Benes, F. M., Heckers, S., et al. (2008). Circuit-based framework for understanding neurotransmitter and risk gene interactions in schizophrenia. *Trends in Neurosciences*, *31*(5), 234–242.

Luby, E. D., Cohen, B. D., Rosenbaum, G., Gottlieb, J. S., & Kelley, R. (1959). Study of a new schizophrenomimetic drug; sernyl.. *AMA Archives of Neurology and Psychiatry*, *81*(3), 363–369.

Millar, J. K., Wilson-Annan, J. C., Anderson, S., Christie, S., Taylor, M. S., Semple, C. A., et al. (2000). Disruption of two novel genes by a translocation co-segregating with schizophrenia. *Human Molecular Genetics*, *9*(9), 1415–1423.

Miyakawa, T, Leiter, L. M., Gerber, D. J., Gainetdinov, R. R., Sotnikova, T. D., et al. (2003). Conditional calcineurin knockout mice exhibit multiple abnormal behaviors related to schizophrenia. *Proceedings of the National Academy of Sciences of the United States of America*, *100*(15), 8987–8992.

Mohn, A. R., Gainetdinov, R. R., Caron, M. G., & Koller, B. H. (1999). Mice with reduced NMDA receptor expression display behaviors related to schizophrenia. *Cell*, *98*(4), 427–436.

Monaghan, D. T., Bridges, R. J., & Cotman, C. W. (1989). The excitatory amino acid receptors: Their classes, pharmacology and distinct properties in the function of the central nervous system. *Annual Review of Pharmacology and Toxicology*, *29*, 365–402.

Moy, S. S., Nadler, J. J., Perez, A., Barbaro, R. P., Johns, J. M., Magnuson, T. R., et al. (2004). Sociability and preference for social novelty in five inbred strains: An approach to assess autistic-like behavior in mice. *Genes, Brain and Behavior*, *3* (5), 287–302.

Nilsson, M., Waters, S., Waters, N., Carlsson, A., & Carlsson, M. L. (2001). A behavioural pattern analysis of hypoglutamatergic mice — effects of four different antipsychotic agents. *Journal of Neural Transmission*, *108*(10), 1181–1196.

Pilowsky, L. S., Bressan, R. A., Stone, J. M., Erlandsson, K., Mulligan, R. S., Krystal, J. H., et al. (2006). First in vivo evidence of an NMDA receptor deficit in medication-free schizophrenic patients. *Molecular Psychiatry*, *11*(2), 118–119.

Ramsey, A. J., Laakso, A., Cyr, M., Sotnikova, T. D., Salahpour, A., Medvedev, I. O., et al. (2008). Genetic NMDA receptor deficiency disrupts acute and chronic effects of cocaine but not amphetamine. *Neuropsychopharmacology*, *33*(11), 2701–2714.

Siapas, A. G., Lubenov, E. V., & Wilson, M. A. (2005). Prefrontal phase locking to hippocampal theta oscillations. *Neuron*, *46*(1), 141–151.

Sirota, A., Montgomery, S., Fujisawa, S., Isomura, Y., Zugaro, M., & Buzsaki, G. (2008). Entrainment of neocortical neurons and gamma oscillations by the hippocampal theta rhythm.. *Neuron*, *60*(4), 683–697.

Stefansson, H., Sigurdsson, E., Steinthorsdottir, V., Bjornsdottir, S., Sigmundsson, T., Ghosh, S., et al. (2002). Neuregulin 1 and susceptibility to schizophrenia. *American Journal of Human Genetics*, *71*(4), 877–892.

Walsh, T., McClellan, J. M., McCarthy, S. E., Addington, A. M., Pierce, S. B., Cooper, G. M., et al. (2008). Rare structural variants disrupt multiple genes in neurodevelopmental pathways in schizophrenia.. *Science*, *320*(5875), 539–543.

Winterer, G., Ziller, M., Dorn, H., Frick, K., Mulert, C., Wuebben, Y., et al. (2000). Schizophrenia: Reduced signal-to-noise ratio and impaired phase-locking during information processing. *Clinical Neurophysiology*, *111*(5), 837–849.

Yeragani, V. K., Cashmere, D., Miewald, J., Tancer, M., & Keshavan, M. S. (2006). Decreased coherence in higher frequency ranges (beta and gamma) between central and frontal EEG in patients with schizophrenia: A preliminary report. *Psychiatry Research*, *141*(1), 53–60.

A. Sawa (Ed.)
Progress in Brain Research, Vol. 179
ISSN 0079-6123

CHAPTER 7

Modeling excess striatal D2 receptors in mice

Christoph Kellendonk*

Department of Psychiatry, Department of Pharmacology, New York, NY, USA

Abstract: Dopaminergic hyperactivity in the striatum has been one of the most replicated physiological findings in patients with schizophrenia. To study the consequences of increased D2 receptor density in the striatum, as it has been observed in patients with schizophrenia, D2 receptors have been selectively overexpressed in the mouse striatum. Here, the analysis of this mouse model is summarized and discussed in the context of the dopamine hypothesis of schizophrenia.

Keywords: schizophrenia; dopamine hypothesis; D2 receptors; transgenic mouse; striatum

The dopamine hypothesis of schizophrenia

The dopamine hypothesis of schizophrenia has so far been the most influential hypothesis about schizophrenia (Howes and Kapur, 2009). In 1966 Jacques Van Rossum proposed that "overstimulation of dopamine receptors could be part of the etiology" of schizophrenia (for a historical review: (Baumeister and Francis, 2002)). The hypothesis was originally based on the observation that known psycho-stimulants, such as amphetamine, induce stereotypic motor behaviors. These behaviors could be blocked by antipsychotic medication, such as chlorpromazine, which by interfering with dopamine function was known to lead to parkinsonian-like movement disorders. Arguably, the strongest support for the dopamine hypothesis was provided in the 1970th by Solomon Snyder and Philip Seeman who found that the efficacy of antipsychotic medication correlated directly with its occupancy of dopamine receptors.

The identification of an effective drug target for psychosis does however not necessarily imply that this target needs to be involved in the pathophysiology or even the etiology of schizophrenia. One initial criticism of the dopamine hypothesis has therefore been that it is not based on measurable physiological alterations in the dopamine system. Because overwhelming evidence for alterations in the brain dopamine system has been found in the last two decades, a role of dopamine in the pathophysiology of the disease is not questioned any longer by most scientists. Whether dopamine is also involved in the etiology of the disease is still unknown. This question is much harder to address because schizophrenia is considered a neuro-developmental disease, consequently patients are diagnosed long after the disease has started its course.

Dopaminergic hyperfunction in the striatum

A dopaminergic hyperfunction in schizophrenia, as originally postulated, has been found in the striatum of patients using a variety of techniques. Some, though not all, postmortem studies

*Corresponding author.
Tel: +1 212 342 3114; Fax: +1 212 305 8780;
E-mail: ck491@columbia.edu

DOI: 10.1016/S0079-6123(09)17907-4

have detected, an increase in subcortical dopamine and homo-vanillic acid (HVA: a metabolite of dopamine) even in unmedicated patients (Davis et al., 1991). This is consistent with the repeated findings of increased Fluoro-Dopa (F-Dopa) uptake in the striatum of patients with schizophrenia, which measures dopamine synthesis by the dopamine synthesizing enzyme L-amino-acid decarboxylase (Frankle, 2007). An increase in F-Dopa uptake has recently been found also in prodromal patients (Howes et al., 2009). Because prodromal patients have not yet fully developed the disease, this is an indication of increased striatal dopamine synthesis at an early stage in the disease process. Evidence for elevated dopamine release in the striatum comes from PET imaging studies showing an augmented amphetamine-induced displacement of D2 receptor binding in the striatum of patients with schizophrenia (Frankle, 2007).

These alterations in presynaptic dopamine availability are paralleled by changes in striatal dopamine receptors. An upregulation of striatal D2 receptors was originally observed in postmortem studies of drug-free patients and later confirmed with PET imaging studies (Davis et al., 1991; Laruelle, 1998). However, not all imaging studies have found increased densities in D2 receptors, most likely due to the heterogeneous nature of the disease. A meta-analysis performed in 1998 by Marc Laruelle calculated a 12% increase in striatal D2 receptor density after comparing 13 imaging studies (Laruelle, 1998). Antipsychotic medication also increases D2 receptor density, but the elevated D2 receptor binding potential has been observed in drug-free and drug-naive patients (Laruelle, 1998). Higher dopamine availability as measured by amphetamine-induced dopamine release should lead to an increased occupancy of D2 receptors in the striatum, which was later found by Anissa Abi-Dargham and colleagues (Abi-Dargham et al., 2000). Moreover, elevated occupancy of D2 receptors was predictive for a good treatment response of positive symptoms with antipsychotic drugs (Abi-Dargham et al., 2000). Because elevated occupancy of D2 receptors correlates with amphetamine-induced dopamine release in patients with schizophrenia, both measurements may identify the same pathological alteration (Abi-Dargham et al., 2009).

Cortical dopaminergic hypofunction

Since depletion of dopamine in the prefrontal cortex impairs working memory in monkeys (Brozoski et al., 1979) and working memory deficits are central to schizophrenia, it has been thought that altered dopamine input to the prefrontal cortex could be involved in the cognitive deficits of schizophrenia. Rather than a hyperfunction of the prefrontal dopamine system, as measured in the striatum and postulated by the original hypothesis of schizophrenia, a hypofunction of the prefrontal dopamine system was proposed (Davis et al., 1991; Weinberger, 1987). Direct evidence for a decrease in presynaptic dopamine synthesis was later provided by a postmortem study that showed a decrease in tyrosine hydroxylase immunoreactivity in the prefrontal cortex (Akil et al., 1999). If in schizophrenia the cortical dopaminergic projections are indeed hypoactive, D1 receptors may be upregulated in the cortex of patients, to compensate for decrease in dopamine release. Using [11C]NNC as a D1-specific PET ligand, an increase in D1 receptor density has been found in the prefrontal cortex of patients with schizophrenia. This increase correlates with deficits in working memory suggesting that the upregulation is functionally linked to the cognitive deficits (Abi-Dargham et al., 2002), but see conflicting results of (Okubo et al., 1997).

In conclusion, brain imaging and postmortem studies have found strong evidence for a dopaminergic hyperfunction in the striatum of patients with schizophrenia that includes an increase in the release of dopamine and an increase in the density and occupancy of the D2 receptor. This subcortical dopaminergic hyperfunction is thought to contribute to the positive symptoms. They also have found less extensive evidence for a dopaminergic hypofunction in the cortex that may contribute to the cognitive symptoms.

Modeling increased density of D2 receptors in the striatum of the mouse

Increased striatal D2 receptor activation may account for the positive symptoms that can be treated with selective D2 receptor antagonists. Nevertheless, a causal relationship between increased D2 receptor activation and the appearance of symptoms is difficult to draw, simply because too many other factors are changed in the brain of patients with schizophrenia. To directly study the behavioral consequences of increased striatal D2 receptor density that has been observed in the imaging studies, my colleagues Eleanor Simpson, Eric Kandel and I generated mice in which we over-expressed D2 receptors selectively in the striatum (D2R-OE mice) (Kellendonk et al., 2006). Obviously, it is impossible to model schizophrenia in its entirety in the mouse including all symptoms clusters: the positive, negative and cognitive symptoms. The positive symptoms are particularly hard to model because it is not possible to measure hallucinations or delusions in mice. We therefore focused in this mouse model on studying specific endophenotypes of the disease that are related to the cognitive and negative symptoms and can be studied in mice (for a full discussion of modeling endophenotypes of schizophrenia in mice see (Kellendonk et al. (2009). To combine striatal selectivity with temporal control of transgenic D2 receptor expression, we used an artificial transactivator system — the tetracycline transactivator (tTA) system (Fig. 1).

Thirty percent of striatal cells express transgenic receptors in D2R-OE mice. Expression is restricted to the postsynaptic medium spiny neurons. Transgenic D2 receptors are neither expressed in cholinergic interneurons nor in the presynaptic dopaminergic midbrain neurons that project to the striatum (Drew et al., 2007). Saturation ligand binding studies using the D2 receptor antagonist [3H]-methyl-spiperone show that double transgenic mice have 15% higher receptor binding capacity than their littermates, similar to the calculated 12% increase observed with imaging studies in schizophrenic patients (Laruelle, 1998). Moreover, dopamine-induced adenylate cyclase activity is reduced in striatal membrane preparations of D2R-OE mice, consistent with increased coupling of D2R to Gi/o proteins, an inhibitor of adenylate cyclase (Kellendonk et al., 2006).

To study the behavioral consequences of selective upregulation of D2 receptors in the striatum, D2R-OE were analyzed in a battery of behavioral tasks (Table 1). D2R-OE mice exhibit specific deficits in cognitive tasks that are sensitive to medial prefrontal lesions in the mouse, including working memory and conditional associative learning (CAL) (Bach et al., 2008; Kellendonk et al., 2006; Ward et al., 2009). Working memory and CAL are impaired in patients with schizophrenia and require the frontal cortex not only in mice, but also in humans, non-human primates and rats (Goldman-Rakic, 1994; Weinberger and Berman, 1996). The cognitive deficits of D2R-OE mice are specific because not all forms of learning are affected. Spatial reference memory is not affected when assayed in a spatial T- as well as in the water maze. Two forms of learned fear, contextual and cued fear conditioning, are also not affected like the learning of simple stimulus response tasks (Table 1). In addition, D2R-OE mice show normal anxiety levels and no deficits in sensory-motor gating (Table 1). In conclusion, D2R-OE mice display selective deficits in cognitive tasks that are sensitive to prefrontal lesions.

How do excess striatal D2 receptors induce deficits in prefrontal-dependent tasks? The striatum participates in several cortico-striatal-pallido-thalamo-cortical loops (Haber, 2003). Disruption at any component of these loops could lead to deficits in prefrontal-dependent behaviors. Conceivably therefore, the cognitive impairments in D2R-OE mice are due to the effects of excess D2 receptors on the efficiency of the cortico-striatal synapse. Alternatively, excess striatal D2 receptors could indirectly affect prefrontal cortical function. In line with the second hypothesis dopamine tissue levels, turnover and D1 receptor activation are altered in the prefrontal cortex of D2R-OE mice (Kellendonk et al., 2006). Numerous studies have shown in rodents, non-human primates and humans that prefrontal D1 receptors modulate working memory with an inverted

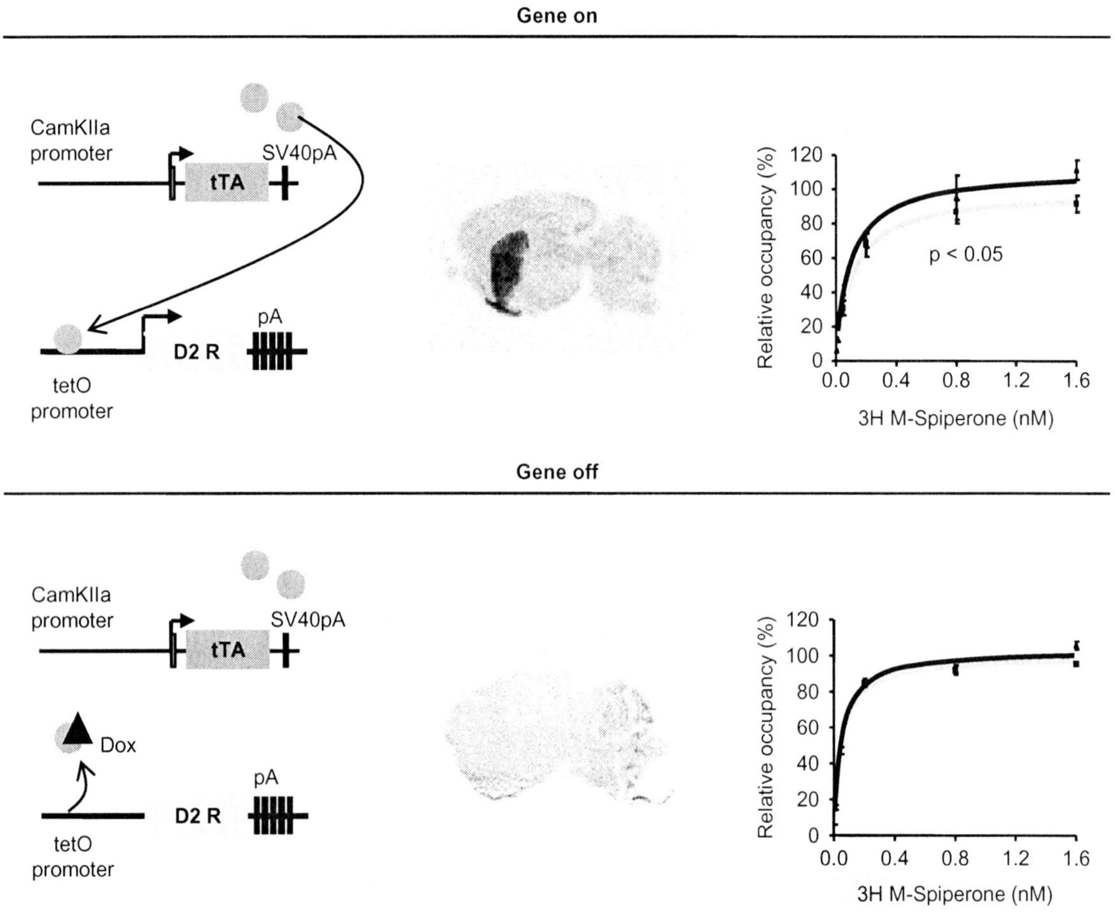

Fig. 1. Reversible and selective over-expression of dopamine D2 receptors in the striatum. Top: Transgenic D2 receptor expression in the absence of doxycycline: "Gene on". Top left: The tetracycline transactivator, tTA, is expressed under the control of the CamKIIalpha gene promoter. In the absence of the antibiotikum doxycycline, tTA binds to the artificial tetracycline response element (tetO) promoter and drives the expression of transgenic D2 receptors (SV40pA, pA: polyadenylation signals required for proper expression of the transgene). Top middle: A sagital brain slice depicting mRNA expression of transgenic D2 receptors in a mouse carrying both transgenes, the CamKIIalpha-tTA transgene and the tetO-D2R transgene. Only transgenic D2 receptor mRNAs are detected in this in situ hybridization. Transgene expression is almost completely restricted to the striatum. Top right: A saturating ligand binding study shows a 15% increase in total (endogenous plus transgenic) D2 receptors in striatal membranes (D2R-OE mice: black; control mice: gray). Bottom: In the "Gene off" condition when mice are fed with doxycycline transgene expression is switched off and D2 receptor protein levels are reversed to control levels (adapted from Kellendonk et al., 2006).

U-shaped function, in which too much or too little D1 activation leads to deficits (Vijayraghavan et al., 2007). In consequence, alterations in D1 receptor signaling in the prefrontal cortex of D2R-OE mice could contribute to their cognitive deficits.

Schizophrenia has a developmental component and it is possible that developmental upregulation of D2Rs is sufficient to induce the behavioral deficits. In line with this switching-off the expression of excess D2 receptors in the adult animal fails to reverse the deficits in the working memory T-maze and CAL tasks (Bach et al., 2008; Kellendonk et al., 2006). Moreover, even when the transgene is switched off at birth, adult D2R-OE mice are impaired in performing the T-maze working memory task. This suggests that over-expression of D2Rs during early

Table 1. Behavioral analysis of D2R-OE mice

Behavior:	Task	Performance		Reference
		Gene on adult	Gene off adult	
Working memory	Delayed non-match to sample T-maze	Impaired	Deficit persists[a]	Kellendonk et al. (2006)
	Delayed win-shift 8 arm radial arm maze	Impaired		Kellendonk et al. (2006)
Working memory/ sustained attention	Operant conditioning based peak timing task:			
	Bisection procedure with long intervals (12 vs. 24 s)	Impaired		Ward et al. (2009)
Reversal learning	Odor discrimination reversal task: Reversal of rule	Impaired[c]		Kellendonk et al. (2006)
Spatial reference memory	Water maze: hidden platform	Normal		Kellendonk et al. (2006)
	Spatial T-maze	Normal		Kellendonk et al. (2006)
Learned fear	Contextual fear conditioning	Normal		*Unpublished*
	Cued fear conditioning	Normal		*Unpublished*
Conditioned associative learning	Operant conditioning based CAL task	Impaired	Deficit persists	Bach et al. (2008)
Simple learning and discrimination	Acquisition of operant learning	Normal		Ward et al. (2009)
	Simple tone discrimination	Normal		*Unpublished*
	Water maze: visible platform	Normal		Kellendonk et al. (2006)
	Odor discrimination reversal task: acquisition	Normal		Kellendonk et al. (2006)
Interval timing	Operant conditioning based peak timing task:	Impaired	Deficits partially reversed[b]	Drew et al. (2007)
	Bisection procedure with short intervals (2 vs. 8 s)	Normal		Ward et al. (2009)
Appetitive incentive motivation	Operant conditioning based progressive ratio task	Impaired	Deficit reversed	Drew et al. (2007)
Social interaction	Social interaction with ovarectomized female	Impaired	Deficit persists	*Unpublished*
Locomotor activity	Open field: locomotor activity	Normal[d]		Kellendonk et al. (2006)
Anxiety	Open field: time and activity in center	Normal		Kellendonk et al. (2006)
	Elevated plus maze	Normal		Kellendonk et al. (2006)
Sensory-motor gating	Startle box: pre-pulse inhibition	Normal		Kellendonk et al. (2006)
	Acoustic startle reflex	Normal		Kellendonk et al. (2006)

[a]Deficit persists even after switching off the transgene at birth.
[b]The deficit was only partially reversed because accuracy was reversed but not increased variability.
[c]The impairment was only evident when looking at latencies but not accuracy.
[d]D2R-OE mice are less active in the first 20 min (of 1 h) of open field testing.

development is indeed sufficient to cause cognitive deficits in the adult mouse.

In schizophrenia, the severities of cognitive and negative symptoms are highly correlated, leading to the hypothesis that cognitive and negative symptoms are causally related. The negative symptoms include apathy, anhedonia and a decrease in the anticipation of expected reward. Using an operant, progressive ratio schedule to measure motivation, we found that D2R-OE mice showed significant impairment while the transgene was switched on, but behaved normally after the transgene was switched off, suggesting that dysfunction of the D2Rs in the striatum directly contributes to negative-like symptoms in the mouse (Drew et al., 2007; Ward et al., 2009).

Concluding remarks

In the context of schizophrenia, the findings obtained with the D2R-OE mice suggest four important new ideas: First, striatal D2 receptors may not only be involved in the generation of positive symptoms but also in the generation of the cognitive symptoms. Second, the prefrontal hypofunction that is observed in patients with schizophrenia is not necessarily of primary origin but could be a consequence of dopaminergic hyperfunction in the striatum. This course of events would be the opposite of what is commonly believed: that the subcortical dopaminergic hyperactivity is a consequence of the cortical abnormalities (Meyer-Lindenberg et al., 2002). Third, one reason why pharmacological blockade of D2 receptors has minimal, if any beneficial effects on the cognitive symptoms of schizophrenia is because antipsychotic medication is given too late. Over-activation of D2 receptors early during development leads to irreversible changes in the brain that cannot be reversed anymore by treatment with antipsychotic medication during adolescence or early adulthood. Fourth, the cognitive and negative symptoms may share some etiological components, which is consistent with the observation that the severity of these two types of symptoms strongly correlates in patients.

The key question with regard to the dopamine hypothesis of schizophrenia is whether dopamine is not only part of the pathophysiology of the disorder but also involved in its etiology. Related to this question is how early in the course of the disease dopamine is acting. Association studies have found common allelic variants in the dopamine system associated with the disorder (Talkowski et al., 2008). For example in a subpopulation of patients, altered D2 receptor binding may be genetically determined (Betcheva et al., 2009; Lawford et al., 2005). In other patients, genetic variants may indirectly affect the function of dopamine gene products (Iizuka et al., 2007). However, for most patients, over-activation of the subcortical dopamine system may reflect the limited capacity of the brain to react to early developmental impacts. This idea is based on the observation that a variety of animal models in which development has been disrupted pre- or early postnatally show a hyperactive subcortical dopamine system. This observation is independent of whether the inducing mechanism is a viral infection, toxic agent or brain lesion. The mammalian brain therefore seems to have a limited capacity to compensate against developmental disturbances and one way to compensate for a disrupted development is to activate the subcortical dopamine system. In this sense, dopamine may indeed be the "final common pathway" (Howes and Kapur, 2009).

References

Abi-Dargham, A., Mawlawi, O., Lombardo, I., Gil, R., Martinez, D., Huang, Y., et al. (2002). Prefrontal dopamine D1 receptors and working memory in schizophrenia. *Journal of Neuroscience*, *22*, 3708–3719.

Abi-Dargham, A., Rodenhiser, J., Printz, D., Zea-Ponce, Y., Gil, R., Kegeles, L. S., et al. (2000). Increased baseline occupancy of D2 receptors by dopamine in schizophrenia [comment]. *Proceedings of the National Academy of Sciences of the United States of America*, *97*, 8104–8109.

Abi-Dargham, A., van de Giessen, E., Slifstein, M., Kegeles, L. S., & Laruelle, M. (2009). Baseline and amphetamine-stimulated dopamine activity are related in drug-naive schizophrenic subjects. *Biological Psychiatry*, *65*, 1091–1093.

Akil, M., Pierri, J. N., Whitehead, R. E., Edgar, C. L., Mohila, C., Sampson, A. R., et al. (1999). Lamina-specific alterations in the dopamine innervation of the prefrontal cortex in schizophrenic subjects. *American Journal of Psychiatry*, *156*, 1580–1589.

Bach, M. E., Simpson, E. H., Kahn, L., Marshall, J. J., Kandel, E. R., & Kellendonk, C. (2008). Transient and selective overexpression of D2 receptors in the striatum causes persistent deficits in conditional associative learning. *Proceedings of the National Academy of Sciences of the United States of America*, *105*, 16027–16032.

Baumeister, A. A., & Francis, J. L. (2002). Historical development of the dopamine hypothesis of schizophrenia. *Journal of the History of Neurosciences*, *11*, 265–277.

Betcheva, E. T., Mushiroda, T., Takahashi, A., Kubo, M., Karachanak, S. K., Zaharieva, I. T., et al. (2009). Case–control association study of 59 candidate genes reveals the DRD2 SNP rs6277 (C957T) as the only susceptibility factor for schizophrenia in the Bulgarian population. *Journal of Human Genetics*, *54*, 98–107.

Brozoski, T. J., Brown, R. M., Rosvold, H. E., & Goldman, P. S. (1979). Cognitive deficit caused by regional depletion of dopamine in prefrontal cortex of rhesus monkey. *Science*, *205*, 929–932.

Davis, K. L., Kahn, R. S., Ko, G., & Davidson, M. (1991). Dopamine in schizophrenia: A review and reconceptualization [comment]. *American Journal of Psychiatry*, *148*, 1474–1486.

Drew, M. R., Simpson, E. H., Kellendonk, C., Herzberg, W. G., Lipatova, O., Fairhurst, S., et al. (2007). Transient overexpression of striatal D2 receptors impairs operant motivation and interval timing. *Journal of Neuroscience*, *27*, 7731–7739.

Frankle, W. G. (2007). Neuroreceptor imaging studies in schizophrenia. *Harvard Review of Psychiatry*, *15*, 212–232.

Goldman-Rakic, P. S. (1994). Working memory dysfunction in schizophrenia. *Journal of Neuropsychiatry & Clinical Neurosciences*, *6*, 348–357.

Haber, S. N. (2003). The primate basal ganglia: Parallel and integrative networks. *Journal of Chemical Neuroanatomy*, *26*, 317–330.

Howes, O. D., & Kapur, S. (2009). The dopamine hypothesis of schizophrenia: Version III — the final common pathway. *Schizophrenia Bulletin*, *35*, 549–562.

Howes, O. D., Montgomery, A. J., Asselin, M. C., Murray, R. M., Valli, I., Tabraham, P., et al. (2009). Elevated striatal dopamine function linked to prodromal signs of schizophrenia. *Archives of General Psychiatry*, *66*, 13–20.

Iizuka, Y., Sei, Y., Weinberger, D. R., & Straub, R. E. (2007). Evidence that the BLOC-1 protein dysbindin modulates dopamine D2 receptor internalization and signaling but not D1 internalization. *Journal of Neuroscience*, *27*, 12390–12395.

Kellendonk, C., Simpson, E. H., & Kandel, E. R. (2009). Modeling cognitive endophenotypes of schizophrenia in mice. *Trends in Neuroscience*, *32*, 347–358.

Kellendonk, C., Simpson, E. H., Polan, H. J., Malleret, G., Vronskaya, S., Winiger, V., et al. (2006). Transient and selective overexpression of dopamine D2 receptors in the striatum causes persistent abnormalities in prefrontal cortex functioning.[comment]. *Neuron*, *49*, 603–615.

Laruelle, M. (1998). Imaging dopamine transmission in schizophrenia. A review and meta-analysis. *Quarterly Journal of Nuclear Medicine*, *42*, 211–221.

Lawford, B. R., Young, R. M., Swagell, C. D., Barnes, M., Burton, S. C., Ward, W. K., et al. (2005). The C/C genotype of the C957T polymorphism of the dopamine D2 receptor is associated with schizophrenia. *Schizophrenia Research*, *73*, 31–37.

Meyer-Lindenberg, A., Miletich, R. S., Kohn, P. D., Esposito, G., Carson, R. E., Quarantelli, M., et al. (2002). Reduced prefrontal activity predicts exaggerated striatal dopaminergic function in schizophrenia. *Nature Neuroscience*, *5*, 267–271.

Okubo, Y., Suhara, T., Suzuki, K., Kobayashi, K., Inoue, O., et al. (1997). Decreased prefrontal dopamine D1 receptors in schizophrenia revealed by PET [comment]. *Nature*, *385*, 634–636.

Talkowski, M. E., Kirov, G., Bamne, M., Georgieva, L., Torres, G., Mansour, H., et al. (2008). A network of dopaminergic gene variations implicated as risk factors for schizophrenia. *Human Molecular Genetics*, *17*, 747–758.

Vijayraghavan, S., Wang, M., Birnbaum, S. G., Williams, G. V., & Arnsten, A. F. (2007). Inverted-U dopamine D1 receptor actions on prefrontal neurons engaged in working memory. *Nature Neuroscience*, *10*, 376–384.

Ward, R., Kellendonk, C., Simpson, E. H., Lipatova, O., Drew, M. R., Fairhurst, S., et al. (2009) Impaired timing precision produced by striatal D2 receptor overexpression is mediated by cognitive and motivational deficits. Behavioral Neuroscience, *123*, 720–730.

Weinberger, D. R. (1987). Implications of normal brain development for the pathogenesis of schizophrenia. *Archives of General Psychiatry*, *44*, 660–669.

Weinberger, D. R., & Berman, K. F. (1996). Prefrontal function in schizophrenia: Confounds and controversies. *Philosophical Transactions of the Royal Society of London — Series B: Biological Sciences*, *351*, 1495–1503.

A. Sawa (Ed.)
Progress in Brain Research, Vol. 179
ISSN 0079-6123

CHAPTER 8

Differential function of phosphodiesterase families in the brain: gaining insights through the use of genetically modified animals

Michele P. Kelly* and Nicholas J. Brandon

Wyeth Research, Department of Neuroscience, NJ, USA

Abstract: Phosphodiesterases (PDEs) are the only known enzymes to degrade cAMP and cGMP, intracellular signaling molecules key to numerous cellular functions. There are 11 PDE families identified to date, and each is expressed in a unique pattern across brain regions. Here, we review genetic mouse models in which PDEs are either directly manipulated (e.g., genetically deleted) or are changed in a compensatory manner due to the manipulation of another target. We believe these genetic mouse models have contributed to our understanding of how the PDE1, PDE4, and PDE10 families contribute uniquely to overall brain function.

Keywords: cAMP; cGMP; cyclic nucleotide; second messenger; mouse model; transgenic; knockout; mutant; psychiatry; schizophrenia; phosphodiesterase; intracellular signaling

Phosphodiesterases (PDEs) are the only enzymes known to degrade cyclic nucleotides, and there are 11 families of PDEs identified to date (c.f., Bender and Beavo, 2006; Lugnier, 2006). Most PDE families are expressed in the brain, but no two PDEs show the same regional/subcellular distribution, enabling exquisitely precise regulation of cyclic nucleotide signaling (c.f., Houslay and Milligan, 1997). Great attention has been paid in recent years to the potential of PDEs as therapeutic targets for treatment of neurological dysfunction and psychiatric disorders, particularly schizophrenia (Houslay et al., 2005; Brandon et al., 2007; Menniti et al., 2007). As such, it is important to gain a full understanding of how each PDE family influences brain function. Genetically modified mice are an important tool for gaining such an understanding, and we review here mouse models published to date in which the expression of various PDEs is either directly or indirectly altered.

PDE1 mice

The dual-specificity (meaning degrades both cAMP and cGMP) PDE1 family is comprised of three gene products, PDE1A, PDE1B, and PDE1C, which are differentially expressed across brain regions. PDE1A shows the greatest expression in CA1–3 of hippocampus, the amygdala, and throughout the

*Corresponding author.
Tel.: +1 732 274 4101; Fax: +1 732 274 4755;
E-mail: kellym31@wyeth.com

DOI: 10.1016/S0079-6123(09)17908-6

deep layers of cortex (Yan et al., 1994). In contrast, PDE1B is particularly enriched in the striatum, dentate gyrus of hippocampus, the olfactory bulb and tubercle, and throughout all layers of cortex (Polli and Kincaid, 1994; Yan et al., 1994). PDE1C shows a more restricted pattern of expression, with high levels in the cerebellum and low levels in the striatum and outer layers of the olfactory bulb (Yan et al., 1995).

PDE1B−/− male and female mice (N3 on C57BL6 background) exhibit increased basal and methamphetamine-stimulated locomotor activity as well as impaired spatial learning and memory (Reed et al., 2002). Siuciak et al. (2007a) also showed that PDE1B−/− male mice (N10 on a C57BL/6N CRL background) exhibit increased basal and methamphetamine-stimulated locomotor activity as well as increased hyperactivity in response to amphetamine and the glutamatergic antagonists MK-801 (males and females) and PCP (males only). In contrast, PDE1B−/− mice show normal behaviors related to pain perception (hot plate), anxiety (elevated plus maze), depression (forced swim test), and conditioning (passive and conditioned avoidance; Siuciak et al., 2007a). The behavioral alterations observed in PDE1B−/− mice may be related to increased levels of serotonin in the hippocampus and/or increased levels of dopamine, cAMP-stimulated phospho-DARPP-32 (Thr34), and/or phospho-GluR1(Ser845) in striatum (Reed et al., 2002; Siuciak et al., 2007a; but see Ehrman et al., 2006). Indeed, deleting both PDE1B and DARPP-32 appears to attenuate the heightened locomotor response to methamphetamine observed in PDE1B−/− mice as well as the increased anxiety-related behavior (elevated zero maze) observed in DARPP-32−/− mice (Ehrman et al., 2006). That said, PDE1B/DARPP-32−/− mice show increased basal levels of locomotor activity, impaired visual cue learning (cued water maze), and deficits in behavioral flexibility (reversal learning in Moris Water Maze) relative to PDE1B−/− mice and DARPP-32−/− mice (Ehrman et al., 2006).

Characterization of PDE1C−/− mice has been limited to date; however, data in these mice suggests that PDE1C may play a role in regulating transduction in olfactory sensory neurons (Cygnar and Zhao, 2009). This is particularly interesting in the context of schizophrenia because patients are known to exhibit olfactory deficits that may be linked, in particular, to dysfunction of cAMP signaling (Turetsky and Moberg, 2009).

PDE4 mice

The PDE4 family is, arguably, the most intensely investigated PDE family, particularly in the context of psychiatric and neurological disorders (Houslay et al., 2005). The PDE4 family is cAMP-specific and comprised of four genes, PDE4A–D. PDE4A, 4B, and 4D appear to be expressed relatively ubiquitously throughout the rodent brain with only subtle differences in distribution across brain regions, while PDE4C may only be expressed in the olfactory bulb (Engels et al., 1995; Iona et al., 1998; Takahashi et al., 1999; Perez-Torres et al., 2000). The PDE4 family is dynamically regulated in a brain region-specific manner following the administration of psychoactive agents (Takahashi et al., 1999; Dlaboga et al., 2006, 2008). In addition, the administration of PDE4 inhibitors to rodents alters behaviors relevant to schizophrenia (O'Donnell and Frith, 1999; Zhang et al., 2000; Barad et al., 2004; Houslay et al., 2005; Kanes et al., 2007; Kelly et al., 2007, 2009; Siuciak et al., 2007b).

PDE4B is of particular interest to the field of psychiatry because it has been genetically linked to schizophrenia and has been shown to interact with the schizophrenia candidate gene disrupted in schizophrenia 1 (Millar et al., 2005; Murdoch et al., 2007; Pickard et al., 2007; Fatemi et al., 2008). When on a mixed C57BL6 × 129/OLA background (Zhang et al., 2008), PDE4B−/− male mice exhibit increased anxiety-related behaviors across a number of paradigms (light/dark box, holeboard, and open field) as well as a selective antidepressant-related behavior (in forced swim but not tail suspension). It is possible that the anxiogenic and antidepressant-related phenotypes observed in PDE4B−/− mice are related to increases in plasma corticosterone and hippocampal cell proliferation, respectively (Zhang et al., 2008). Despite these alterations,

PDE4B−/− mice on a mixed background profile normally in tests of pain threshold (hot plate), basal locomotor activity, and learning and memory (water maze; passive avoidance). When the line is on a C57BL/6N background (N8), male PDE4B−/− mice continue to show normal pain threshold (hot plate), intact learning and memory (water maze, passive avoidance; acquisition of conditioned avoidance responding), and a trend to an antidepressant-like phenotype in the forced swim test (Siuciak et al., 2008a). In contrast, however, PDE4B−/− mice on a C57BL/6N background diverge in that they show no increase in anxiety-related behaviors in an elevated plus maze and show a decrease in basal locomotor activity (Siuciak et al., 2007b, 2008a). In addition, these PDE4B−/− mice exhibit increased auditory startle, decreased sensorimotor gating (prepulse inhibition (PPI) of acoustic startle), as well as a slight increase in their sensitivity to the indirect dopaminergic agonist amphetamine (Siuciak et al., 2008a). These results stand in contrast to reports showing that acute administration of the pan-PDE4 inhibitor rolipram decreases anxiety-related behaviors on the elevated plus maze (Silvestre et al., 1999), increases sensorimotor gating, and blocks the effects of amphetamine (Kanes et al., 2007; Kelly et al., 2007, 2009). The behavioral phenotypes observed in PDE4B−/− mice on the C57BL/6N background (N8) may be related to alterations in the dopaminergic and/or serotonergic system as decreases in dopamine turnover are found in striatum and frontal cortex and decreases in serotonin turnover are found in striatum, frontal cortex, and hippocampus of these mice (Siuciak et al., 2008a).

PDE4D−/− male mice (mixed C57BL6 × 129/OLA background) have also been behaviorally characterized. Although PDE4D−/− mice are normal on a test of anxiety-related behavior (elevated plus maze), these mice show robust antidepressant-like phenotypes on both the forced swim and tail suspension tests (Zhang et al., 2002). Interestingly, PDE4D−/− mice show reduced antidepressant-like effects of rolipram but no reduction in effects of the antidepressants desipramine and fluoxetine (Zhang et al., 2002). In addition to mood-related phenotypes, these PDE4D−/− mice also exhibit deficits in long-term, but not short-term, associative memory (cued and contextual fear conditioning; Rutten et al., 2008). Further, although basal synaptic transmission remains normal, PDE4D−/− mice exhibit enhancements in select form of long-term potentiation (e.g., theta burst; Rutten et al., 2008). Together, these reports suggest that PDE4B may play a greater role in anxiety-related behaviors, sensorimotor gating, and dopamine signaling, whereas PDE4D may play a greater role in depressive-related behavior and memory consolidation.

PDE10 mice

The dual-specificity PDE10 family is encoded by one gene, PDE10A, and has emerged as a key therapeutic target for schizophrenia. Interest in PDE10A relates to its unique distribution in rodent brain, which is very high in the caudate putamen, nucleus accumbens, and the olfactory tubercle, and minimal in cortex, hippocampus, and cerebellum. PDE10A−/− mice were originally published on a pure DBA1Lac/J background (PDE10−/−^{DBA}, males and females; Siuciak et al., 2006) and later on a C57BL/6N background (PDE10−/−^{C57}, N10 males only; Siuciak et al., 2008b). On both backgrounds, PDE10A−/− mice show a decrease in basal locomotor activity, impaired learning of a conditioned avoidance responding task, and decreased sensitivity to the hyperlocomotor effects of the glutamatergic antagonist MK-801. In contrast, only the PDE10A−/−^{DBA} mice show decreased effects of the glutamatergic antagonist PCP and only the PDE10A−/−^{C57} mice exhibit increased sensitivity to the hyperlocomotor effects of the indirect dopaminergic agonists amphetamine and methamphetamine. The latter sensitivity observed in PDE10A−/−^{C57} mice may be related to the increased dopamine turnover and/or increased serotonin levels observed in striatum of PDE10−/−^{C57} mice. Phenotypes observed in PDE10A−/− mice are largely consistent with effects of PDE10A inhibitors, with the exception that PDE10A inhibitors are shown to block the effects of dopaminergic agonists (Grauer et al., in press).

A second PDE10A mutant mouse, reportedly targeting only the PDE10A2 spliceform, has also been characterized (Sano et al., 2008). PDE10A2−/− mice (N10 on C57BL6 background) show a trend toward decreased basal locomotor activity, increased anxiety-related behaviors (open field and elevated plus maze, but not light/dark box), and improved motor coordination (rotarod), but these effects do not reach the level of statistical significance. PDE10A2−/− mice also appear normal on assays of pain sensitivity (hot plate), associative learning and memory (contextual and cued fear conditioning), and depressive-related behavior (forced swim test). Interestingly, despite the trend toward increased anxiety-related behavior, PDE10A2−/− mice exhibit a striking increase in sociality (number and durations of social contacts made with a stranger mouse; Sano et al., 2008), consistent with the effect of PDE10A inhibitors (Grauer et al., in press). This difference in sociality cannot be attributed to an increased preference for social odors as both PDE10A2−/− and wild-type littermates exhibit an equivalent preference for a social odor (urine in absence of an actual stimulus mouse) vs. a non-social odor (lemon). Authors of these manuscripts argue that results in PDE10−/− and PDE10A2−/− mice suggest PDE10A inhibitors might prove useful as broad spectrum treatments for schizophrenia, able to treat both positive (e.g., psychoses, delusions, hallucinations) and negative symptoms (e.g., social deficits). Indeed, PDE10 inhibitors have exhibited efficacy in paradigms intended to model the positive, negative, and cognitive symptoms associated with schizophrenia (Grauer et al., in press).

Other mouse models showing changes in PDE activity

In addition to the above mouse models in which the expression of a given PDE isoform has been directly manipulated, there are examples in the literature where PDEs are altered in response to a genetic manipulation in a related signaling pathway. Gαs is the G-protein subunit that stimulates adenylate cyclase, the enzyme that catalyzes the formation of cAMP. Gαs transgenic mice, which overexpress either a wild-type isoform (Gαs mice) or a constitutively active isoform (Gαs* mice) in the forebrain, exhibit a number of behavioral (hyperlocomotor activity, sensorimotor gating deficits, and hippocampus-dependent associative and spatial memory impairments) and anatomical endophenotypes (enlarged ventricle, reduced striatum) associated with psychiatric diseases such as schizophrenia (Kelly et al., 2007, 2008, 2009). These behavioral abnormalities appear to be related to compensatory decreases in cortical and hippocampal cAMP that are triggered by a PKA-dependent upregulation of total PDE activity. Studies in Gαs* mice suggest that this upregulation of total PDE activity is specifically due to increased activity of PDE1 and, possibly, PDE4 (Kelly et al., 2008). Consistent with a role for upregulated PDE activity in the behavioral deficits of these mice, the pan-PDE4 inhibitor rolipram ameliorates sensorimotor gating deficits observed in Gαs and Gαs* mice (Kelly et al., 2007, 2009).

Mice with ENU-induced mutations in the schizophrenia candidate gene Disc1 also show PDE4 alterations (Clapcote et al., 2007). The Disc1 Q31L mutant exhibits depressive-related (forced swim test, tail suspension test, antidepressant responsivity) and schizophrenia-related phenotypes (sensorimotor gating deficit in PPI, working memory deficit, decreased sociality, decreased brain volume). The L100P mutant also shows schizophrenia-related phenotypes (hyperactivity, sensorimotor gating deficits in PPI, working memory deficits, antipsychotic responsivity, decreased brain volume; Clapcote et al., 2007). The overlapping but separable nature of the behavioral phenotypes of these two mutants may be related to alterations in PDE4B. Both the Disc1 Q31L and L100P mutants show decreased binding to PDE4B, in vitro (Clapcote et al., 2007). In vivo, however, only the Q31L mutant exhibits reduced PDE4B activity. Consistent with this observation, the L100P – but not the Q31L – mice show behavioral rescue with rolipram (Clapcote et al., 2007).

Changes in PDEs have also been of particular interest in Huntington's disease mouse models. Both the R6/1 and R6/2 mice, which express a mutant huntingtin gene with an inserted CAG repeat, show a robust reduction in striatal PDE10A mRNA and protein levels, as well as a smaller decrease in PDE1B expression (Hebb et al., 2004; Hu et al., 2004). Interestingly, the reduction in PDE10A expression preceeds onset of motor symptoms. A slight reduction in PDE4A mRNA also occurs in the more severe R6/2 mouse, but only at the early age of 3 weeks when striatal PDE4A mRNA expression is highest (Hebb et al., 2004). Consistent with these observations in rodents, PDE10A protein is decreased in tissue from Huntington's patients (Hebb et al., 2004). It is possible that this downregulation of striatal PDE levels may be an attempt to compensate for the mutant huntingtin, because administration of either PDE4 or PDE10 inhibitors ameliorates neuropathology and behavioral deficits observed in the R6/2 and quinolinic acid model for Huntington's disease (DeMarch et al., 2007, 2008; Giampa et al., 2009a, b).

Conclusion

Studies in genetically modified animal models have greatly contributed to our understanding of how each PDE family contributes uniquely to the function of the nervous system. For example, it is interesting to note that deletion of PDE1B and PDE10A results in nearly opposite phenotypes, despite the fact that both are a dual-specificity PDEs with enriched expression in the striatum. Although our understanding is growing, much work remains. Genetic models are lacking for a great many of the PDE family isoforms and only a single spliceform model exists (PDE10A2−/−). Further, of the genetic models that do exist for PDEs, none allow for temporal control of the genetic manipulation (with the exception of the Gαs transgenic mice described above). Future studies using genetically modified animals will continue to foster our understanding of how PDE families differentially contribute to brain function, which will have important implications for considering the numerous PDE families as therapeutic targets for diseases such as schizophrenia.

References

Barad, M., Blouin, A. M., & Cain, C. K. (2004). Like extinction, latent inhibition of conditioned fear in mice is blocked by systemic inhibition of L-type voltage-gated calcium channels. *Learning & Memory*, *11*, 536–539.

Bender, A. T., & Beavo, J. A. (2006). Cyclic nucleotide phosphodiesterases: Molecular regulation to clinical use. *Pharmacological Reviews*, *58*, 488–520.

Brandon, N. J., & Rotella, D. P. (2007). Potential CNS applications for phosphodiesterase enzyme inhibitors. In J. E. Macor (Ed.), *Annual reports in medicinal chemistry (Vol. 42, Chapter 1)* (pp. 3–12). USA: Elsevier . pp.

Clapcote, S. J., Lipina, T. V., Millar, J. K., Mackie, S., Christie, S., Ogawa, F., et al. (2007). Behavioral phenotypes of Disc1 missense mutations in mice. *Neuron*, *54*, 387–402.

Cygnar, K. D., & Zhao, H. (2009). Phosphodiesterase 1C is dispensable for rapid response termination of olfactory sensory neurons. *Nature Neuroscience*, *12*, 454–462.

DeMarch, Z., Giampa, C., Patassini, S., Bernardi, G., & Fusco, F. R. (2008). Beneficial effects of rolipram in the R6/2 mouse model of Huntington's disease. *Neurobiology of Disease*, *30*, 375–387.

DeMarch, Z., Giampa, C., Patassini, S., Martorana, A., Bernardi, G., & Fusco, F. R. (2007). Beneficial effects of rolipram in a quinolinic acid model of striatal excitotoxicity. *Neurobiology of Disease*, *25*, 266–273.

Dlaboga, D., Hajjhussein, H., & O'Donnell, J. M. (2006). Regulation of phosphodiesterase-4 (PDE4) expression in mouse brain by repeated antidepressant treatment: Comparison with rolipram. *Brain Research*, *1096*, 104–112.

Dlaboga, D., Hajjhussein, H., & O'Donnell, J. M. (2008). Chronic haloperidol and clozapine produce different patterns of effects on phosphodiesterase-1B, −4B, and −10A expression in rat striatum. *Neuropharmacology*, *54*, 745–754.

Ehrman, L. A., Williams, M. T., Schaefer, T. L., Gudelsky, G.A., Reed, T. M., Fienberg, A. A., et al. (2006). Phosphodiesterase 1B differentially modulates the effects of methamphetamine on locomotor activity and spatial learning through DARPP32-dependent pathways: Evidence from PDE1B-DARPP32 double-knockout mice. *Genes, Brain, & Behavior*, *5*, 540–551.

Engels, P., Abdel'Al, S., Hulley, P., & Lubbert, H. (1995). Brain distribution of four rat homologues of the Drosophila dunce cAMP phosphodiesterase. *Journal of Neuroscience Research*, *41*, 169–178.

Fatemi, S. H., King, D. P., Reutiman, T. J., Folsom, T. D., Laurence, J. A., Lee, S., et al. (2008). PDE4B polymorphisms and decreased PDE4B expression are associated with schizophrenia. *Schizophrenia Research*, *101*, 36–49.

Giampa, C., Middei, S., Patassini, S., Borreca, A., Marullo, F., Laurenti, D., et al. (2009a). Phosphodiesterase type IV inhibition prevents sequestration of CREB binding protein, protects striatal parvalbumin interneurons and rescues motor deficits in the R6/2 mouse model of Huntington's disease. *European Journal of Neuroscience*, *29*, 902–910.

Giampa, C., Patassini, S., Borreca, A., Laurenti, D., Marullo, F., Bernardi, G., et al. (. Phosphodiesterase 10 inhibition reduces striatal excitotoxicity in the quinolinic acid model of Huntington's disease. *Neurobiology of Disease*, *34*, 450–456.

Grauer S., Pulito P., Navarra R, Kelly MP, Kelley C., Graf R., et al. (in press) PDE10A inhibitor activity in preclinical models of the positive, cognitive, and negative symptoms of schizophrenia. *JPET*. Epub ahead of print.

Hebb, A. L., Robertson, H. A., & Denovan-Wright, E. M. (2004). Striatal phosphodiesterase mRNA and protein levels are reduced in Huntington's disease transgenic mice prior to the onset of motor symptoms. *Neuroscience*, *123*, 967–981.

Houslay, M. D., & Milligan, G. (1997). Tailoring cAMP-signalling responses through isoform multiplicity. *Trends in Biochemical Sciences*, *22*, 217–224.

Houslay, M. D., Schafer, P., & Zhang, K. Y. (2005). Keynote review: Phosphodiesterase-4 as a therapeutic target. *Drug Discovery Today*, *10*, 1503–1519.

Hu, H., McCaw, E. A., Hebb, A. L., Gomez, G. T., & Denovan-Wright, E. M. (2004). Mutant huntingtin affects the rate of transcription of striatum-specific isoforms of phosphodiesterase 10A. *European Journal of Neuroscience*, *20*, 3351–3363.

Iona, S., Cuomo, M., Bushnik, T., Naro, F., Sette, C., Hess, M., et al. (1998). Characterization of the rolipram-sensitive, cyclic AMP-specific phosphodiesterases: Identification and differential expression of immunologically distinct forms in the rat brain. *Molecular Pharmacology*, *53*, 23–32.

Kanes, S. J., Tokarczyk, J., Siegel, S. J., Bilker, W., Abel, T., & Kelly, M. P. (2007). Rolipram: A specific phosphodiesterase 4 inhibitor with potential antipsychotic activity. *Neuroscience*, *144*, 239–246.

Kelly, M. P., Cheung, Y. F., Favilla, C., Siegel, S. J., Kanes, S.J., Houslay, M. D., et al. (2008). Constitutive activation of the G-protein subunit Galphas within forebrain neurons causes PKA-dependent alterations in fear conditioning and cortical Arc mRNA expression. *Learning & Memory*, *15*, 75–83.

Kelly, M. P., Isiegas, C., Cheung, Y. F., Tokarczyk, J., Yang, X., Esposito, M. F., et al. (2007). Constitutive activation of Galphas within forebrain neurons causes deficits in sensorimotor gating because of PKA-dependent decreases in cAMP. *Neuropsychopharmacology*, *32*, 577–588.

Kelly, M. P., Stein, J. M., Vecsey, C. G., Favilla, C., Yang, X., Bizily, S. F., et al. (2009). Developmental etiology for neuroanatomical and cognitive deficits in mice overexpressing Galphas, a G-protein subunit genetically linked to schizophrenia. *Molecular Psychiatry*, *14*, 398–415.

Lugnier, C. (2006). Cyclic nucleotide phosphodiesterase (PDE) superfamily: A new target for the development of specific therapeutic agents. *Pharmacology & Therapeutics*, *109*, 366–398.

Menniti, F. S., Chappie, T. A., Humphrey, J. M., & Schmidt, C.J. (2007). Phosphodiesterase 10A inhibitors: A novel approach to the treatment of the symptoms of schizophrenia. *Current Opinion in Investigational Drugs*, *8*, 54–59.

Millar, J. K., Pickard, B. S., Mackie, S., James, R., Christie, S., Buchanan, S. R., et al. (2005). DISC1 and PDE4B are interacting genetic factors in schizophrenia that regulate cAMP signaling. *Science*, *310*, 1187–1191.

Murdoch, H., Mackie, S., Collins, D. M., Hill, E. V., Bolger, G.B., Klussmann, E., et al. (2007). Isoform-selective susceptibility of DISC1/phosphodiesterase-4 complexes to dissociation by elevated intracellular cAMP levels. *Journal of Neuroscience*, *27*, 9513–9524.

O'Donnell, J. M., & Frith, S. (1999). Behavioral effects of family-selective inhibitors of cyclic nucleotide phosphodiesterases. *Pharmacology, Biochemistry & Behavior*, *63*, 185–192.

Perez-Torres, S., Miro, X., Palacios, J. M., Cortes, R., Puigdomenech, P., & Mengod, G. (2000). Phosphodiesterase type 4 isozymes expression in human brain examined by in situ hybridization histochemistry and[3H]rolipram binding autoradiography. Comparison with monkey and rat brain. *Journal of Chemical Neuroanatomy*, *20*, 349–374.

Pickard, B. S., Thomson, P. A., Christoforou, A., Evans, K. L., Morris, S. W., Porteous, D. J., et al. (2007). The PDE4B gene confers sex-specific protection against schizophrenia. *Psychiatric Genetics*, *17*, 129–133.

Polli, J. W., & Kincaid, R. L. (1994). Expression of a calmodulin-dependent phosphodiesterase isoform (PDE1B1) correlates with brain regions having extensive dopaminergic innervation. *Journal of Neuroscience*, *14*, 1251–1261.

Reed, T. M., Repaske, D. R., Snyder, G. L., Greengard, P., & Vorhees, C. V. (2002). Phosphodiesterase 1B knock-out mice exhibit exaggerated locomotor hyperactivity and DARPP-32 phosphorylation in response to dopamine agonists and display impaired spatial learning. *Journal of Neuroscience*, *22*, 5188–5197.

Rutten, K., Misner, D. L., Works, M., Blokland, A., Novak, T.J., Santarelli, L., et al. (2008). Enhanced long-term potentiation and impaired learning in phosphodiesterase 4D-knockout (PDE4D) mice. *European Journal of Neuroscience*, *28*, 625–632.

Sano, H., Nagai, Y., Miyakawa, T., Shigemoto, R., & Yokoi, M. (2008). Increased social interaction in mice deficient of the striatal medium spiny neuron-specific phosphodiesterase 10A2. *Journal of Neurochemistry*, *105*, 546–556.

Silvestre, J. S., Fernandez, A. G., & Palacios, J. M. (1999). Effects of rolipram on the elevated plus-maze test in rats: A preliminary study. *Journal of Psychopharmacology*, *13*, 274–277.

Siuciak, J. A., Chapin, D. S., McCarthy, S. A., & Martin, A. N. (2007a). Antipsychotic profile of rolipram: Efficacy in rats and reduced sensitivity in mice deficient in the phosphodiesterase-4B (PDE4B) enzyme. *Psychopharmacology*, *192*, 415–424.

Siuciak, J. A., McCarthy, S. A., Chapin, D. S., Fujiwara, R. A., James, L. C., Williams, R. D., et al. (2006). Genetic deletion of the striatum-enriched phosphodiesterase PDE10A: Evidence for altered striatal function. *Neuropharmacology*, *51*, 374–385.

Siuciak, J. A., McCarthy, S. A., Chapin, D. S., & Martin, A. N. (2008a). Behavioral and neurochemical characterization of mice deficient in the phosphodiesterase-4B (PDE4B) enzyme. *Psychopharmacology, 197*, 115–126.

Siuciak, J. A., McCarthy, S. A., Chapin, D. S., Martin, A. N., Harms, J. F., & Schmidt, C. J. (2008b). Behavioral characterization of mice deficient in the phosphodiesterase-10A (PDE10A) enzyme on a C57/Bl6N congenic background. *Neuropharmacology, 54*, 417–427.

Siuciak, J. A., McCarthy, S. A., Chapin, D. S., Reed, T. M., Vorhees, C. V., & Repaske, D. R. (2007b). Behavioral and neurochemical characterization of mice deficient in the phosphodiesterase-1B (PDE1B) enzyme. *Neuropharmacology, 53*, 113–124.

Takahashi, M., Terwilliger, R., Lane, C., Mezes, P. S., Conti, M., & Duman, R. S. (1999). Chronic antidepressant administration increases the expression of cAMP-specific phosphodiesterase 4A and 4B isoforms. *Journal of Neuroscience, 19*, 610–618.

Turetsky, B. I., & Moberg, P. J. (2009). An odor-specific threshold deficit implicates abnormal intracellular cyclic AMP signaling in schizophrenia. *American Journal of Psychiatry, 166*, 226–233.

Yan, C., Bentley, J. K., Sonnenburg, W. K., & Beavo, J. A. (1994). Differential expression of the 61 kDa and 63 kDa calmodulin-dependent phosphodiesterases in the mouse brain. *Journal of Neuroscience, 14*, 973–984.

Yan, C., Zhao, A. Z., Bentley, J. K., Loughney, K., Ferguson, K., & Beavo, J. A. (1995). Molecular cloning and characterization of a calmodulin-dependent phosphodiesterase enriched in olfactory sensory neurons. *Proceedings of the National Academy of Sciences of the United States of America, 92*, 9677–9681.

Zhang, H. T., Crissman, A. M., Dorairaj, N. R., Chandler, L. J., & O'Donnell, J. M. (2000). Inhibition of cyclic AMP phosphodiesterase (PDE4) reverses memory deficits associated with NMDA receptor antagonism. *Neuropsychopharmacology, 23*, 198–204.

Zhang, H. T., Huang, Y., Jin, S. L., Frith, S. A., Suvarna, N., Conti, M., et al. (2002). Antidepressant-like profile and reduced sensitivity to rolipram in mice deficient in the PDE4D phosphodiesterase enzyme. *Neuropsychopharmacology, 27*, 587–595.

Zhang, H. T., Huang, Y., Masood, A., Stolinski, L. R., Li, Y., Zhang, L., et al. (2008). Anxiogenic-like behavioral phenotype of mice deficient in phosphodiesterase 4B (PDE4B). *Neuropsychopharmacology, 33*, 1611–1623.

A. Sawa (Ed.)
Progress in Brain Research, Vol. 179
ISSN 0079-6123

CHAPTER 9

Gene models of schizophrenia: DISC1 mouse models

Hanna Jaaro-Peled*

Department of Psychiatry and Behavioral Sciences, Johns Hopkins University School of Medicine, Baltimore, MD, USA

Abstract: *Disrupted in Schizophrenia-1* (*DISC1*) is one of the most likely susceptibility genes for schizophrenia (SZ). DISC1 is being established as a hub protein with various functions in the pre- and postnatal development of the nervous system. Since generation of a knockout (KO) mouse has proved challenging, various alternative approaches have been taken. Seven DISC1 mouse models have been described to date. All of them display neuroanatomical and behavioral abnormalities relevant to SZ, although most of them have not been fully characterized yet, requiring further analysis.

NRG1 and ErbB4, also highly promising susceptibility genes for SZ, share many features with DISC1. They are involved in various aspects of pre- and postnatal neurodevelopment. The NRG1 and ErbB4 mouse models also display neuroanatomical and behavioral abnormalities similar to the DISC1 mouse models.

In the future, four main directions need further study. First, further characterization of the seven DISC1 mouse models, especially in light of basic research findings. Second, more extensive employment of the inducible models. Third, generation of a DISC1 KO. Fourth, combination of the DISC1 mouse models with other risk factors: crossing with other genetic models such as NRG1/ErbB4 mutants and exposure to environmental risk factors.

Keywords: schizophrenia; DISC1; NRG1; ErbB4; mouse models; neuroanatomy; behavior

Introduction

Disrupted in Schizophrenia-1 (*DISC1*) was identified in a Scottish pedigree where its disruption by a balanced t(1;11) translocation segregates with major mental illness including schizophrenia (SZ) (Millar et al., 2000; St Clair et al., 1990). In this pedigree the translocation is causal, although not fully penetrant (Blackwood et al., 2001). *DISC1* has since been shown by linkage and association studies to be involved in major mental illness in various ethnic groups, making it a very promising susceptibility gene, although no other causative mutations have been identified (Chubb et al., 2008). The identification of DISC1 prompted great interest in its then totally unknown biological functions. Today it is known that DISC1 carries out multiple functions in the nervous system by interacting with various proteins in different cell compartments from embryonic development until adulthood (Jaaro-Peled et al., 2009). Therefore, mutations or even variations in DISC1 have the potential to affect a variety of

*Corresponding author.
Tel.: +1 41 061 41780; Fax: +1 41 061 41792;
E-mail: hjaarop1@jhmi.edu

DOI: 10.1016/S0079-6123(09)17909-8

neural processes and be manifest as a disorder (Porteous et al., 2006). Many mouse models of DISC1 have been described. Since a knockout (KO) has proved difficult to generate, probably due to the not well-understood multi-isomeric nature of DISC1 (Ma et al., 2002), various alternative approaches have been taken.

This review will start with a short introduction about DISC1, followed by a brief description of each DISC1 mouse model, focusing on general (Table 1) and behavioral (Table 2) comparisons among them.

The many roles of DISC1

Basic research is gradually unraveling the many roles of DISC1. The major ones will be discussed here, since they are of obvious relevance for characterization of the DISC1 mouse models.

DISC1 inhibits GSK3β by direct interaction thereby regulating the proliferation of neuronal progenitor cells (Mao et al., 2009). As a component of the dynein motor complex together with Lissencephaly-1 (LIS1) and Nuclear Distribution Element-Like-1 (NDEL1), DISC1 is crucial for radial migration of the pyramidal neurons in the developing cortex (Kamiya et al., 2005). Interestingly, DISC1 interacts also with the kinesin-1 motor protein, regulates the localization of NUDEL/LIS1/14-3-3ε complex into the axons, and thus regulates axon elongation (Taya et al., 2007). A significant pool of DISC1 is localized to the nucleus, where it interacts with the CREB transcription factor family members ATF4/5 (Morris et al., 2003; Sawamura et al., 2008). Another pool of DISC1 is localized in the synapse (Kirkpatrick et al., 2006) and interacts with synaptic proteins (Camargo et al., 2007). In the adult brain DISC1 is expressed mainly in the dentate gyrus, where it regulates integration of new neurons (Duan et al., 2007) and presynaptic development (Faulkner et al., 2008). Another binding partner of DISC1 is phosphodiesterase 4 (PDE4B) (Millar et al., 2005), suggesting a role in cAMP signaling.

Mouse DISC1 is only ~60% identical to the human DISC1 on both the nucleotide and the amino acid levels (Ma et al., 2002). This relatively rapid evolution of DISC1 is an important feature to take into account when discussing DISC1 mouse models.

Seven DISC1 mouse models

Seven DISC1 mouse models have been generated to date. For an excellent scheme of the genetic designs of the DISC1 mouse models, please see the recent review by (Kellendonk et al. (2009).

Spontaneous mutation in DISC1 in the 129 strain (Δ25 bp mutant)

In the process of a gene targeting experiment with *DISC1*, a 25 bp spontaneous deletion was identified in exon 6 in the 129S6/SvEv mouse strain. The deletion induces a frameshift that creates a premature stop codon in exon 7 (Koike et al., 2006). This does not seem to affect DISC1 mRNA level (Koike et al., 2006), but it is not clear what happens at the protein level (Ishizuka et al., 2007). This mutant mouse *DISC1* was transferred to C57BL/6J background for further characterization (Koike et al., 2006; Kvajo et al., 2008).

ENU-induced missense DISC1 mutants

N-nitroso-*N*-ethylurea (ENU) induces point mutations at a high locus-specific rate. One can screen the gene of interest for mutations. Screening for missense mutations in exon 2 of DISC1 yielded two independent missense mutants: Q31L and L100P. Interestingly, each of the missense mutations in DISC1 is sufficient to elicit a variety of anatomical, biochemical, and behavioral abnormalities. Even more impressive, each mutant displays distinct behavioral abnormalities: Q31L was proposed as depression-like and L100P as SZ-like (Clapcote et al., 2007).

Constitutive and inducible αCaMKII-ΔC DISC1 transgenic (CaMK-ΔC Tg)

αCaMKII (α-calmodulin kinase II) has been widely used in neurobiological research to drive

Table 1. Comparison among the DISC1 mouse models

	Mutants			Dominant-negative transgenic			
				Constitutive		Inducible	
References	Clapcote et al. (2007)		Koike et al. (2006), Kvajo et al. (2008)	Shen et al. (2008)	Hikida et al. (2007)	Pletnikov et al. (2008)	Li et al. (2007)
Name	Q31L	L100P	Δ25 bp	BAC-ΔC	CaMK-ΔC	CaMK-ΔC	CaMK-cc
Exogenous DISC1	n/a	n/a	n/a	Mouse	Human	Human	Human
Promoter	Endogenous	Endogenous	Endogenous	Endogenous	αCaMKII	αCaMKII	αCaMKII
Mouse strain	B6 × DBA → B6		129 → B6	CBA × B6	B6	B6 × SJL × CBA	B6
Neuroanatomy	Reduced brain volume	Reduced brain volume	Reduced PFC volume	Smaller cortex, enlarged LV, corpus callosum agenesis, thinning of layers II/III	Enlarged LV	Enlarged LV	
Neurogenesis				Reduced			
Pyramidal cells			DG: altered organization, CA1: reduced short-term potentiation				DG: reduced dendritic complexity, CA1: reduced synaptic transmission
Interneurons			No difference in PV or calbindin in mPFC	PV reduced in mPFC, hippocampus	PV reduced in mPFC		

CA1, cornu ammonis; DG, dentate gyrus; LV, lateral ventricles; mPFC, medial prefrontal cortex; PV, parvalbumin.

Table 2. Behavioral analysis of the DISC1 mouse models

Putative neuropsychological correlate	Test	Mutants			Dominant-negative transgenic			
					Constitutive		Inducible	
		Clapcote et al. (2007)		Koike et al. (2006)	Shen et al. (2008)	Hikida et al. (2007)	Pletnikov et al. (2008)	Li et al. (2007)
		Q31L	L100P	Δ25 bp	BAC-ΔC	CaMK-ΔC	CaMK-ΔC	CaMK-cc
Cognition	Prepulse inhibition	yes	yes	no	n/d	yes	no	n/d
	Latent inhibition	yes	yes	n/d	yes	n/d	n/d	n/d
	Working memory (T maze)	yes	yes	yes	n/d	n/d	n/d	yes
	Spatial learning and memory	no	no	no	n/d	no	yes (only F)	n/d
Positive	Novelty-induced locomotion (first 30 min in open field)	no	yes	no	no	no	no	no
Negative or depressive-like	Sociability (three-chamber)	yes	no	n/d	n/d	n/d	n/d	yes
	Sucrose consumption	yes	no	n/d	n/d	n/d	n/d	n/d
	Forced swim test	yes	no	n/d	yes	yes	n/d	yes
Gender used for testing		M+F	M+F	M+F	M	M	M+F	M+F

Abnormal phenotype in the test is indicated by "yes", no difference from the controls by "no", and not determined yet by "n/d". The gender of the mice used for behavioral testing is indicated as: F, females; M, males.

expression of transgenes postnatally in the pyramidal neurons in the forebrain (Mayford et al., 1996). Owing to the high expression of endogenous DISC1 in the cortex and hippocampus and the importance of the forebrain in SZ, there are three transgenic DISC1 models that utilize this promoter to drive the expression of putative dominant-negative transgenes. Human *DISC1* truncated as in the Scottish pedigree in which DISC1 was originally detected (corresponding to amino acids 1-597) was expressed under the αCaMKII promoter (CaMK-ΔC Tg). Constitutive (Hikida et al., 2007) and conditional Tet-off double transgenic system (Pletnikov et al., 2008) versions have been generated. Notably, the first publication about the inducible CaMK-ΔC Tg did not utilize the inducible feature of the system. Evidence showing that the C′-terminally truncated DISC1 (ΔC) acts in a dominant-negative manner has been obtained both in culture and in vivo (Sawa and Snyder, 2005).

αCaMKII-cc DISC1 transgenic (CaMK-cc Tg)

The third transgenic model that utilized the αCaMKII promoter and a dominant-negative principle expresses the C'-terminal portion of human DISC1 (amino acids 671-852) competing with the endogenous wild-type DISC1 on binding to the important DISC1 interactors, NUDEL and LIS1. Expression of the transgene is induced by tamoxifen for less than 2 days. In the published paper the authors activated the dominant-negative transgene transiently at postnatal day 7, which remarkably was sufficient to elicit significant abnormalities evident in adulthood, whereas adult induction was not (Li et al., 2007).

BAC-ΔC DISC1 transgenic (BAC-ΔC Tg)

The newest DISC1 mouse model overexpresses genomic DNA of truncated mouse *DISC1* encoding the first 8 exons by using a bacterial artificial chromosome (BAC). In comparison with the transgenic models described above (using the ectopic CaMK promoter and human DISC1), this approach has the advantages of an endogenous promoter and potentially more complete inhibition of the endogenous wild-type mouse DISC1 (Shen et al., 2008).

Changes in gross anatomy in DISC1 mouse models (Table 1)

Imaging studies of SZ patients have detected enlarged ventricles accompanied by volume decreases of various brain areas (Vita et al., 2006). DISC1 mouse models have similar gross abnormalities in brain structure. Translational magnetic resonance imaging (MRI) studies have detected reduction in brain volume (specifically cortex, entorhinal cortex, thalamus, and cerebellum) in the Q31L and L100P mutants (Clapcote et al., 2007) and enlarged lateral ventricles (without a change in brain volume) in the CaMK-ΔC Tg (Hikida et al., 2007; Pletnikov et al., 2008). BAC-ΔC Tg is the only model in which both reduced cerebral cortex and enlarged lateral ventricles have been found by histology (Shen et al., 2008). The reduction in the thickness of the cortex was attributed to thinning of layers II/III. A unique finding in these mice was partial agenesis of the corpus callosum, which is expected to interfere with the communication between the two hemispheres. Impaired connectivity of the corpus callosum has also been found in SZ (Miyata et al., 2007).

Cellular and histological abnormalities in DISC1 mouse models

Several types of cellular and histological changes have been reported in brains from patients with SZ. Very interestingly, similar abnormalities are found in brains of DISC1 mouse models.

Neurogenesis in DISC1 mouse models (Table 1)

BAC-ΔC Tg had reduced neuronal proliferation in midneurogenesis, which may be the cause of the thinned cortical layer II/III mentioned above. In light of the recent finding that DISC1 regulates the proliferation of neuronal progenitor cells (Mao et al., 2009), it would be of interest to investigate this aspect also in the other DISC1 models.

Pyramidal cells synaptic structure and function in DISC1 mouse models (Table 1)

The hippocampus is the main region of DISC1 expression in adult mice (Austin et al., 2004). Accordingly, cytoarchitectural abnormalities have been described mainly in the hippocampus, but also in the prefrontal cortex of DISC1 models. Dendritic complexity of dentate gyrus granule cells was reduced and basal synaptic transmission was reduced when the DISC1-cc dominant-negative transgene was activated at postnatal day 7 (P7), but cornu ammonis 1 (CA1) long-term potentiation was normal (Li et al., 2007). In the Δ25 bp mutant, abnormalities were detected selectively in the dentate gyrus. There was a higher proportion of immature neurons in the outer layers of the granule cell layer accompanied with reduced total number, and misorientation of the apical dendrites. Mature granule cells also had misoriented apical dendrites, decrease in total dendrite length, and reduced spine number. Accordingly, short-term potentiation was reduced in the Δ25 bp mutant CA1. Much milder abnormalities were detected in the medial prefrontal cortex in the form of shorter apical dendrites of layer V pyramidal neurons (Kvajo et al., 2008). Similar dendritic abnormalities have also been reported in SZ (Black et al., 2004; Glantz and Lewis, 2000).

Interneurons of DISC1 mouse models (Table 1)

Decrease in immunoreactivity of parvalbumin-positive interneurons is regarded as an important hallmark for SZ. Dysfunction of the fast-spiking interneurons expressing parvalbumin would impair their ability to synchronize firing of pyramidal neurons, contributing to the cognitive dysfunction in SZ (Lewis et al., 2005). Consistently, reduced numbers of parvalbumin-positive interneurons have been detected in the medial prefrontal cortex of both the CaMK-ΔC Tg and the BAC-ΔC Tg and in the hippocampus of the BAC-ΔC Tg (Hikida et al., 2007; Shen et al., 2008). In the BAC-ΔC Tg there was also a difference in the pattern of parvalbumin staining in the dorsolateral prefrontal cortex: an increase in the number of parvalbumin cells in the outer layers at the expense of the inner layers (Shen et al., 2008). In contrast, no differences were found in the numbers of calbindin- or parvalbumin-positive interneurons in the medial prefrontal cortex of the Δ25 bp mutant (Kvajo et al., 2008).

Biochemical abnormalities in DISC1 mouse models

The DISC1–PDE4B connection has been investigated only in the missense mutants. When DISC1 and PDE4B were coexpressed in cell culture, the Q31L and especially the L100P DISC1 mutants had reduced binding to PDE4B. Only the Q31L mutant had lower PDE4B activity, consistent with the resistance of Q31L mutants to the PDE4 inhibitor, rolipram (Clapcote et al., 2007).

Expression of the inducible form of CaMK-ΔC Tg was found to cause transient (at P7, but not at P21) molecular changes: reduced protein levels of its interactors, endogenous mouse DISC1 and LIS1 (but not NDEL1) and also of the presynaptic marker SNAP-25 (but not the postsynaptic marker PSD-95) (Pletnikov et al., 2008).

Neurite outgrowth in primary cultures from DISC1 mouse models

DISC1 is important for neurite outgrowth as shown both by cell culture experiments (Miyoshi et al., 2003; Ozeki et al., 2003) and DISC1 RNAi in primary cultures (Kamiya et al., 2005). Primary cortical neuron cultures from the DISC1 mouse models, inducible CaMK-ΔC Tg (Pletnikov et al., 2008) and BAC-ΔC Tg (Shen et al., 2008), consistently show attenuated neurite outgrowth.

Behavioral analysis of DISC1 mouse models (Table 2)

SZ symptoms have been described to consist of cognitive, "positive" (psychotic), and "negative" (blunted emotional expression, apathy, social withdrawal) components. These neuropsychological domains can be modeled in mice with varying degrees of validity (Arguello and Gogos, 2006). Sensorimotor gating is a neurophysiological process that is essentially the same in rodents and humans, making it an excellent endophenotype (Swerdlow et al., 1999). It can be assessed by the prepulse inhibition (PPI) test. PPI deficit is detected in patients with a variety of neurological disorders, including SZ (Geyer, 2006). A similar deficit has been described in the Q31L and L100P mutants (Clapcote et al., 2007) and in the constitutive CaMK-ΔC Tg (although it is much milder in the latter) (Hikida et al., 2007). Another measure of attention is latent inhibition, which is also impaired in SZ and can be assessed in mouse models (Lubow, 2005). Q31L and L100P mutants (Clapcote et al., 2007) and BAC-ΔC Tg (Shen et al., 2008) were subjected to this test and all of them were impaired. Other cognitive tests are harder to correlate with human tests. The T maze (delayed non-match to place) test has detected abnormalities in the working memory of all DISC1 models to which it has been applied: Δ25 bp mutant (Koike et al., 2006; Kvajo et al., 2008), Q31L and L100P mutants (Clapcote et al., 2007), and CaMK-Δc Tg (Li et al., 2007). Spatial learning and memory as assessed by the Morris water maze test seemed normal in the Δ25 bp (Kvajo et al., 2008), Q31L and L100P mutants (Clapcote et al., 2007), and CaMK-ΔC Tg (Hikida et al., 2007). Only the inducible CaMK-ΔC females (but not the males) were impaired in finding the platform in the probe test (Pletnikov et al., 2008).

Hyperactivity in a novel open field is thought to be analogous to psychomotor agitation, a positive symptom in SZ. Significant hyperactivity during the first 30 min in the open field has been found only in the L100P mutants (Clapcote et al., 2007). The constitutive CaMK-ΔC Tg mutants were significantly hyperactive when tested in the open field for 2 h (Hikida et al., 2007) and the inducible CaMK-ΔC Tg (only males) were hyperactive when tested for 22 h (Pletnikov et al., 2008), but at least the latter is not novelty-induced. Supersensitivity to psychostimulants is another positive symptom of SZ. Thus far there have been no publications on psychostimulant administration to the DISC1 models.

It may be even harder to model negative symptoms of SZ in mice and it is especially unclear how to differentiate between tests relevant to negative symptoms and tests relevant to depression. Since DISC1 is a candidate gene not only for SZ, but also for other mental disorders including depression, such phenotypes are of interest even if not clearly attributed to a specific disease. Social withdrawal can be modeled by a three-chamber sociability test. It has been applied to the Q31L and L100P point mutants (Clapcote et al., 2007) and to the inducible CaMK-Δc Tg (Li et al., 2007) and showed abnormalities in all of them, except for L100P. The constitutive (Hikida et al., 2007) and inducible CaMK-ΔC Tg (Pletnikov et al., 2008) were tested in other social tests, in which only the inducible CaMK-ΔC Tg males differed from the wild-type littermates. Anhedonia can be manifested in mice by decreased reinforcing properties of rewards (Nestler et al., 2002). The Q31L, but not the L100P mutants consume less sucrose in the sucrose consumption test. The forced swim test is widely used to screen for antidepressants, which in general reduce the immobility time in this task (David et al., 2003). The Q31L mutants (but not L100P) (Clapcote et al., 2007), BAC-ΔC Tg (Shen et al., 2008), CaMK-ΔC Tg (Hikida et al., 2007), and CaMK-Δc Tg (Li et al., 2007) displayed behavioral despair (were immobile longer), which may imply a depression-like phenotype or negative-like symptoms. A unique behavior which has been looked at only in the BAC-ΔC Tg is stress calls during stressful conditions (tail suspension test). Interestingly, the BAC-ΔC Tg squeaked significantly less frequently than did their wild-type littermates (Shen et al., 2008).

Experiments to "treat" the behavioral abnormalities in DISC1 mouse models

Experiments to normalize the behavior of DISC1 mouse models with drugs used in psychiatry have been reported only for the L100P and the Q31L mutants (Clapcote et al., 2007). Interestingly, the two missense mutants displayed SZ-like and depression-like profiles, respectively. Antipsychotics partially alleviated the PPI and latent inhibition deficits and the hyperactivity of the L100P mutants, but not the PPI and latent inhibition deficits of the Q31L mutant. The antidepressant bupropion (a dopamine and norepinephrine reuptake inhibitor) was effective at normalizing the mild PPI deficit and the increased forced swim immobility of Q31L, but not at normalizing the PPI deficit of the L100P mutants. The PDE4 inhibitor rolipram normalized the PPI deficit of L100P, but not that of Q31L, consistent with the finding of Q31L having a lower PDE4B activity (see section "Biochemical abnormalities in DISC1 mouse models" above).

Comparison between DISC1 and NRG1/ErbB4 mouse models

Other highly promising susceptibility genes for SZ are *Nrg1* and *ErbB4* (which encodes for the receptor of NRG1) (Harrison and Law, 2006). Both DISC1 (Chubb et al., 2008; Ishizuka et al., 2006; Mackie et al., 2007) and NRG1/ErbB4 (Mei and Xiong, 2008) have important roles in pre- and postnatal development of the nervous system. Evidence for a crosstalk between these two pathways is accumulating (Jaaro-Peled et al., 2009). Various mouse models of NRG1 and ErbB4 have been generated (Falls, 2003). Numerous NRG1 isoforms are generated from the *Nrg1* gene (Buonanno and Fischbach, 2001; Mei and Xiong, 2008), complicating the generation and characterization of the models. Comparison between the phenotypes of the DISC1 and the NRG1/ErbB4 mouse models shows remarkable similarities. Enlarged lateral ventricles and decreased dendritic spine density on subicular pyramidal neurons have been reported in type III Nrg1 KO mice (Chen et al., 2008). Reduced numbers of parvalbumin-positive cells were found in the hippocampus of ErbB4 KO mice (Neddens and Buonanno, 2009) (Table 3).

Behaviorally, PPI deficits have been detected in NRG1 transmembrane domain KO (Stefansson et al., 2002) and type III KO mice (Chen et al., 2008), but not in EGF-like domain KO (Duffy et al., 2008) or Ig-like domain KO (Rimer et al., 2005). Impaired latent inhibition has been demonstrated in an NRG1 Ig-like domain KO (Rimer et al., 2005). The T maze paradigm detected working memory impairment in the NRG1-type III KO mice (Chen et al., 2008). The Morris water maze detected spatial learning and memory abnormalities in the brain-specific ErbB4 KO (Golub et al., 2004). Novelty-induced hyperlocomotion in the open field has been detected in the transmembrane domain and EGF-like domain NRG1 KO (Duffy et al., 2008; O'Tuathaigh et al., 2006). Sociability seemed normal in the transmembrane domain NRG1 KO (O'Tuathaigh et al., 2007). As far as we know, NRG1 and ErbB4 mutant mice have probably not been tested in the forced swim test.

Concluding remarks

All of the DISC1 mouse models display some kind of neuroanatomical or cytoarchitectural abnormalities relevant to SZ. The cytoarchitectural changes have been found mainly in the hippocampal dentate gyrus. Two (CaMK-ΔC Tg, and BAC-ΔC Tg) out of the three models in which this feature was investigated showed reduced parvalbumin immunoreactivity. All of the DISC1 mouse models have behavioral abnormalities relevant to SZ. Most of the models have cognitive abnormalities when subjected to the relevant tests: impaired attention as demonstrated by the PPI and latent inhibition tests and impaired working memory in the delayed non-match to place task. Most of the models have normal spatial learning and memory as judged by the Morris water maze test. Positive symptoms in the form of psychomotor agitation as modeled by hyperactivity during the first 30 min in a novel open field have been demonstrated only in one model (L100P mutant). In the domain of negative/depression-like symptoms, the forced swim test has been most widely applied and showed abnormalities in all except for the L100P mutant. In summary, the L100P mutant seems to be an exception among the DISC1 mouse models.

The differences in behavioral tests between various DISC1 models may be accounted for by the different designs of the models and thus provide valuable information, or by differences in strains (Table 1), in gender (Table 2), and varying methodologies and environments.

Until now, the only use of inducible DISC1 mouse models has been activation of the CaMK-c

Table 3. Comparison of the abnormalities found in the DISC1, NRG1, and ErbB4 mouse models

Feature		Model		
		DISC1	NRG1	ErbB4
Neuroanatomy	Enlarged lateral ventricles	+++ (Hikida et al., 2007; Pletnikov et al., 2008; Shen et al., 2008)	++ (Chen et al., 2008)	n/d
Neurogenesis		++ (Shen et al., 2008)	n/d	n/d
Pyramidal cells		++ (Kvajo et al., 2008; Li et al., 2008)	++ (Chen et al., 2008)	n/d
Interneurons	Reduced parvalbumin+	+++ (Hikida et al., 2007; Shen et al., 2008)	n/d	++ (Neddens and Buonanno 2009)
Behavior				
Cognition	Prepulse inhibition	+++ (Clapcote et al., 2007; Hikida et al., 2007)	++ (Chen et al., 2008; Stefansson et al., 2002)	++ (Stefansson et al., 2002)
	Latent inhibition	+++ (Clapcote et al., 2007; Shen et al., 2008)	++ (Rimer et al., 2005)	n/d
	Working memory (T maze)	+++ (Clapcote et al., 2007; Koike et al., 2006; Kvajo et al., 2008; Li et al., 2007)	++ (Chen et al., 2008)	n/d
	Spatial learning and memory (MWM)	+ (Clapcote et al., 2007)	n/d	++ (Golub et al., 2004)
Positive	Novelty-induced locomotion	+ (Clapcote et al., 2007)	+++ (O'Tuathaigh et al., 2006; Duffy et al., 2008)	+ (O'Tuathaigh et al., 2007; Duffy et al., 2008)
Negative/ depressive-like	Sociability (three-chamber)	+++ (Clapcote et al., 2007; Li et al., 2007)	+ (O'Tuathaigh et al., 2007)	n/d
	Forced swim test	+++ (Clapcote et al., 2007; Hikida et al., 2007; Li et al., 2007; Shen et al., 2008)	n/d	n/d

+++ Tested in ≥3 independent experiments, >50% positive (detected abnormalities).
++ Tested in ≥3 independent experiments, ≤50% positive OR tested in <3 models, >50% positive.
+ Tested in ≥3 independent experiments, <25% positive OR tested in <3 models, ≤50% positive.
MWM, Morris water maze; n/d, not determined.

dominant-negative transgene transiently at P7, which remarkably elicited hippocampal (cytoarchitectural and functional) and behavioral abnormalities in the adults, whereas activation of the transgene in adulthood did not (Li et al., 2007). Further carefully designed use of the inducible models, CaMK-ΔC Tg (Pletnikov et al., 2008) and CaMK-Δc Tg (Li et al., 2007), should help to define the neurodevelopmental time window(s) critical for the emergence of the SZ-like abnormalities that could guide prevention experiments. One limitation is that both inducible DISC1 models employ the CaMK promoter, which is assumed to be functional only after birth (Mayford et al., 1996). Intriguingly, the inducible CaMK-ΔC transgene was reported to be activated from E15 (Pletnikov et al., 2008), which may enable its activation to study the prenatal roles of DISC1.

Theoretically the most straightforward approach to study the functions of DISC1 would be to generate a KO, but it has proven practically difficult. Since the Δ25 bp mutant seems to retain many

DISC1 isoforms (Ishizuka et al., 2007), the generation of a more complete KO is eagerly awaited.

Since SZ is a multifactorial disease, crossing of the DISC1 mouse models with others, such as Nrg1/ErbB4 models or exposure to environmental risk factors, may provide a more realistic model. The constitutive models may be easier to use for these applications.

Acknowledgments

The author would like to thank Dr. Pamela Talalay for critical reading of this manuscript. The author is supported by NARSAD.

Abbreviations

CaMK	Calmodulin kinase
DISC1	Disrupted in schizophrenia-1
KO	knockout
LIS1	Lissencephaly-1
NDEL1	Nuclear Distribution Element-Like-1
Nrg1	Neuregulin-1
SZ	schizophrenia
Tg	transgenic

References

Arguello, P. A., & Gogos, J. A. (2006). Modeling madness in mice: One piece at a time. *Neuron, 52*, 179–196.

Austin, C. P., Ky, B., Ma, L., Morris, J. A., & Shughrue, P. J. (2004). Expression of Disrupted-In-Schizophrenia-1, a schizophrenia-associated gene, is prominent in the mouse hippocampus throughout brain development. *Neuroscience, 124*, 3–10.

Black, J. E., Kodish, I. M., Grossman, A. W., Klintsova, A. Y., Orlovskaya, D., Vostrikov, V., et al. (2004). Pathology of layer V pyramidal neurons in the prefrontal cortex of patients with schizophrenia. *American Journal of Psychiatry, 161*, 742–744.

Blackwood, D. H., Fordyce, A., Walker, M. T., St, Clair, D. M., Porteous, D. J., & Muir, W. J. (2001). Schizophrenia and affective disorders — cosegregation with a translocation at chromosome 1q42 that directly disrupts brain-expressed genes: Clinical and P300 findings in a family. *The American Journal of Human Genetics, 69*, 428–433.

Buonanno, A., & Fischbach, G. D. (2001). Neuregulin and ErbB receptor signaling pathways in the nervous system. *Current Opinion in Neurobiology, 11*, 287–296.

Camargo, L. M., Collura, V., Rain, J. C., Mizuguchi, K., Hermjakob, H., Kerrien, S., Bonnert, T. P., Whiting, P. J., & Brandon, N. J. (2007). Disrupted in Schizophrenia 1 Interactome: Evidence for the close connectivity of risk genes and a potential synaptic basis for schizophrenia. *Molecular Psychiatry, 12*, 74–86.

Chen, Y. J., Johnson, M. A., Lieberman, M. D., Goodchild, R. E., Schobel, S., Lewandowski, N., et al. (2008). Type III neuregulin-1 is required for normal sensorimotor gating, memory-related behaviors, and corticostriatal circuit components. *Journal of Neuroscience, 28*, 6872–6883.

Chubb, J. E., Bradshaw, N. J., Soares, D. C., Porteous, D. J., & Millar, J. K. (2008). The DISC locus in psychiatric illness. *Molecular Psychiatry, 13*, 36–64.

Clapcote, S. J., Lipina, T. V., Millar, J. K., Mackie, S., Christie, S., Ogawa, F., et al. (2007). Behavioral phenotypes of Disc1 missense mutations in mice. *Neuron, 54*, 387–402.

David, D. J., Renard, C. E., Jolliet, P., Hascoet, M., & Bourin, M. (2003). Antidepressant-like effects in various mice strains in the forced swimming test. *Psychopharmacology (Berl), 166*, 373–382.

Duan, X., Chang, J. H., Ge, S., Faulkner, R. L., Kim, J. Y., Kitabatake, Y., et al. (2007). Disrupted-In-Schizophrenia 1 regulates integration of newly generated neurons in the adult brain. *Cell, 130*, 1146–1158.

Duffy, L., Cappas, E., Scimone, A., Schofield, P. R., & Karl, T. (2008). Behavioral profile of a heterozygous mutant mouse model for EGF-like domain neuregulin 1. *Behavioral Neuroscience, 122*, 748–759.

Falls, D. L. (2003). Neuregulins: Functions, forms, and signaling strategies. *Experimental Cell Research, 284*, 14–30.

Faulkner, R. L., Jang, M. H., Liu, X. B., Duan, X., Sailor, K. A., Kim, J. Y., et al. (2008). Development of hippocampal mossy fiber synaptic outputs by new neurons in the adult brain. *Proceedings of the National Academy of Sciences of the United States of America, 105*, 14157–14162.

Geyer, M. A. (2006). The family of sensorimotor gating disorders: Comorbidities or diagnostic overlaps? *Neurotoxicity Research, 10*, 211–220.

Glantz, L. A., & Lewis, D. A. (2000). Decreased dendritic spine density on prefrontal cortical pyramidal neurons in schizophrenia. *Archives of General Psychiatry, 57*, 65–73.

Golub, M. S., Germann, S. L., & Lloyd, K. C. (2004). Behavioral characteristics of a nervous system-specific erbB4 knock-out mouse. *Behavioural Brain Research, 153*, 159–170.

Harrison, P. J., & Law, A. J. (2006). Neuregulin 1 and schizophrenia: Genetics, gene expression, and neurobiology. *Biological Psychiatry, 60*, 132–140.

Hikida, T., Jaaro-Peled, H., Seshadri, S., Oishi, K., Hookway, C., Kong, S., et al. (2007). Dominant-negative DISC1 transgenic mice display schizophrenia-associated phenotypes detected by measures translatable to humans. *Proceedings of the National Academy of Sciences of the United States of America, 104*, 14501–14506.

Ishizuka, K., Chen, J., Taya, S., Li, W., Millar, J. K., Xu, Y., et al. (2007). Evidence that many of the DISC1 isoforms in C57BL/6J mice are also expressed in 129S6/SvEv mice. *Molecular Psychiatry*, *12*, 897–899.

Ishizuka, K., Paek, M., Kamiya, A., & Sawa, A. (2006). A review of Disrupted-In-Schizophrenia-1 (DISC1): Neurodevelopment, cognition, and mental conditions. *Biological Psychiatry*, ***59***, 1189–1197.

Jaaro-Peled, H., Hayashi-Takagi, A., Seshadri, S., Kamiya, A., Brandon, N. J. and Sawa, A. (2009) Neurodevelopmental mechanisms of schizophrenia: Understanding disturbed postnatal brain maturation through neuregulin-1-ErbB4 and DISC1. *Trends in Neuroscience*, *32*, 485–495.

Kamiya, A., Kubo, K., Tomoda, T., Takaki, M., Youn, R., Ozeki, Y., et al. (2005). A schizophrenia-associated mutation of DISC1 perturbs cerebral cortex development. *Nature Cell Biology*, *7*, 1167–1178.

Kellendonk, C., Simpson, E. H., & Kandel, E. R. (2009). Modeling cognitive endophenotypes of schizophrenia in mice. *Trends in Neuroscience*, *32*, 347–358.

Kirkpatrick, B., Xu, L., Cascella, N., Ozeki, Y., Sawa, A., & Roberts, R. C. (2006). DISC1 immunoreactivity at the light and ultrastructural level in the human neocortex. *Journal of Comparative Neurology*, *497*, 436–450.

Koike, H., Arguello, P. A., Kvajo, M., Karayiorgou, M., & Gogos, J. A. (2006). Disc1 is mutated in the 129S6/SvEv strain and modulates working memory in mice. *Proceedings of the National Academy of Sciences of the United States of America*, *103*, 3693–3697.

Kvajo, M., McKellar, H., Arguello, P. A., Drew, L. J., Moore, H., MacDermott, A. B., et al. (2008). A mutation in mouse Disc1 that models a schizophrenia risk allele leads to specific alterations in neuronal architecture and cognition. *Proceedings of the National Academy of Sciences of the United States of America*, *105*, 7076–7081.

Lewis, D. A., Hashimoto, T., & Volk, D. W. (2005). Cortical inhibitory neurons and schizophrenia. *Nature Reviews Neuroscience*, *6*, 312–324.

Li, W., Zhou, Y., Jentsch, J. D., Brown, R. A., Tian, X., Ehninger, D., et al. (2007). Specific developmental disruption of disrupted-in-schizophrenia-1 function results in schizophrenia-related phenotypes in mice. *Proceedings of the National Academy of Sciences of the United States of America*, *104*, 18280–18285.

Lubow, R. E. (2005). Construct validity of the animal latent inhibition model of selective attention deficits in schizophrenia. *Schizophrenia Bulletin*, *31*, 139–153.

Ma, L., Liu, Y., Ky, B., Shughrue, P. J., Austin, C. P., & Morris, J. A. (2002). Cloning and characterization of Disc1, the mouse ortholog of DISC1 (Disrupted-in-Schizophrenia 1). *Genomics*, *80*, 662–672.

Mackie, S., Millar, J. K., & Porteous, D. J. (2007). Role of DISC1 in neural development and schizophrenia. *Current Opinion in Neurobiology*, *17*, 95–102.

Mao, Y., Ge, X., Frank, C. L., Madison, J. M., Koehler, A. N., Doud, M. K., et al. (2009). Disrupted in schizophrenia 1 regulates neuronal progenitor proliferation via modulation of GSK3beta/beta-catenin signaling. *Cell*, *136*, 1017–1031.

Mayford, M., Bach, M. E., Huang, Y. Y., Wang, L., Hawkins, R. D., & Kandel, E. R. (1996). Control of memory formation through regulated expression of a CaMKII transgene. *Science*, *274*, 1678–1683.

Mei, L., & Xiong, W. C. (2008). Neuregulin 1 in neural development, synaptic plasticity and schizophrenia. *Nature Reviews Neuroscience*, *9*, 437–452.

Millar, J. K., Pickard, B. S., Mackie, S., James, R., Christie, S., Buchanan, S. R., et al. (2005). DISC1 and PDE4B are interacting genetic factors in schizophrenia that regulate cAMP signaling. *Science*, *310*, 1187–1191.

Millar, J. K., Wilson-Annan, J. C., Anderson, S., Christie, S., Taylor, M. S., Semple, C. A., et al. (2000). Disruption of two novel genes by a translocation co-segregating with schizophrenia. *Human Molecular Genetics*, *9*, 1415–1423.

Miyata, J., Hirao, K., Namiki, C., Fukuyama, H., Okada, T., Miki, Y., et al. (2007). Interfrontal commissural abnormality in schizophrenia: Tractography-assisted callosal parcellation. *Schizophrenia Research*, *97*, 236–241.

Miyoshi, K., Honda, A., Baba, K., Taniguchi, M., Oono, K., Fujita, T., et al. (2003). Disrupted-In-Schizophrenia 1, a candidate gene for schizophrenia, participates in neurite outgrowth. *Molecular Psychiatry*, *8*, 685–694.

Morris, J. A., Kandpal, G., Ma, L., & Austin, C. P. (2003). DISC1 (Disrupted-In-Schizophrenia 1) is a centrosome-associated protein that interacts with MAP1A, MIPT3, ATF4/5 and NUDEL: Regulation and loss of interaction with mutation. *Human Molecular Genetics*, *12*, 1591–1608.

Neddens, J. and Buonanno, A. (2009) Selective populations of hippocampal interneurons express ErbB4 and their number and distribution is altered in ErbB4 knockout mice. *Hippocampus*.

Nestler, E. J., Gould, E., Manji, H., Buncan, M., Duman, R. S., Greshenfeld, H. K., et al. (2002). Preclinical models: Status of basic research in depression. *Biological Psychiatry*, *52*, 503–528.

O'Tuathaigh, C. M., Babovic, D., O'Sullivan, G. J., Clifford, J. J., Tighe, O., Croke, D. T., et al. (2007). Phenotypic characterization of spatial cognition and social behavior in mice with "knockout" of the schizophrenia risk gene neuregulin 1. *Neuroscience*, *147*, 18–27.

O'Tuathaigh, C. M., O'Sullivan, G. J., Kinsella, A., Harvey, R. P., Tighe, O., Croke, D. T., et al. (2006). Sexually dimorphic changes in the exploratory and habituation profiles of heterozygous neuregulin-1 knockout mice. *Neuroreport*, *17*, 79–83.

Ozeki, Y., Tomoda, T., Kleiderlein, J., Kamiya, A., Bord, L., Fujii, K., et al. (2003). Disrupted-in-Schizophrenia-1 (DISC-1): Mutant truncation prevents binding to NudE-like (NUDEL) and inhibits neurite outgrowth. *Proceedings of the National Academy of Sciences of the United States of America*, *100*, 289–294.

Pletnikov, M. V., Ayhan, Y., Nikolskaia, O., Xu, Y., Ovanesov, M. V., Huang, H., et al. (2008). Inducible expression of mutant human DISC1 in mice is associated with brain and behavioral abnormalities reminiscent of schizophrenia. *Molecular Psychiatry*, *13*, 173–186, 115.

Porteous, D. J., Thomson, P., Brandon, N. J., & Millar, J. K. (2006). The genetics and biology of DISC1 — an emerging role in psychosis and cognition. *Biological Psychiatry*, *60*, 123–131.

Rimer, M., Barrett, D. W., Maldonado, M. A., Vock, V. M., & Gonzalez-Lima, F. (2005). Neuregulin-1 immunoglobulin-like domain mutant mice: Clozapine sensitivity and impaired latent inhibition. *Neuroreport*, *16*, 271–275.

Sawa, A., & Snyder, S. H. (2005). Genetics. Two genes link two distinct psychoses. *Science*, *310*, 1128–1129.

Sawamura, N., Ando, T., Maruyama, Y., Fujimuro, M., Mochizuki, H., Honjo, K., et al. (2008). Nuclear DISC1 regulates CRE-mediated gene transcription and sleep homeostasis in the fruit fly. *Molecular Psychiatry*, *13*, 1138–1148, 1069.

Shen, S., Lang, B., Nakamoto, C., Zhang, F., Pu, J., Kuan, S. L., et al. (2008). Schizophrenia-related neural and behavioral phenotypes in transgenic mice expressing truncated Disc1. *Journal of Neuroscience*, *28*, 10893–10904.

St Clair, D., Blackwood, D., Muir, W., Carothers, A., Walker, M., Spowart, G., et al. (1990). Association within a family of a balanced autosomal translocation with major mental illness. *Lancet*, *336*, 13–16.

Stefansson, H., Sigurdsson, E., Steinthorsdottir, V., Bjornsdottir, S., Sigmundsson, T., Ghosh, S., et al. (2002). Neuregulin 1 and susceptibility to schizophrenia. *American Journal of Human Genetics*, *71*, 877–892.

Swerdlow, N. R., Braff, D. L., & Geyer, M. A. (1999). Cross-species studies of sensorimotor gating of the startle reflex. *Annals of the New York Academy of Sciences*, *877*, 202–216.

Taya, S., Shinoda, T., Tsuboi, D., Asaki, J., Nagai, K., Hikita, T., et al. (2007). DISC1 regulates the transport of the NUDEL/LIS1/14-3-3epsilon complex through kinesin-1. *Journal of Neuroscience*, *27*, 15–26.

Vita, A., De Peri, L., Silenzi, C., & Dieci, M. (2006). Brain morphology in first-episode schizophrenia: A meta-analysis of quantitative magnetic resonance imaging studies. *Schizophrenia Research*, *82*, 75–88.

A. Sawa (Ed.)
Progress in Brain Research, Vol. 179
ISSN 0079-6123

CHAPTER 10

The sandy (*sdy*) mouse: a dysbindin-1 mutant relevant to schizophrenia research

Konrad Talbot*

Center for Neurobiology and Behavior, Department of Psychiatry, University of Pennsylvania, Philadelphia, PA, USA

Abstract: Dysbindin-1 reductions appear to be common in dysfunctional brain areas of schizophrenia cases. In the absence of a dysbindin-1 knockout, sandy (sdy) mice provide our only means of studying the potential contribution of this protein to clinical features of schizophrenia in live animals. Our knowledge of sandy mice is reviewed here. These mice have a deletion mutation that arose spontaneously in DBA/2J mice in the gene encoding dysbindin-1 (*Dtnbp1*). This null protein mutation ($Dtnbp1^{sdy}$) leads to an absence of dysbindin-1 in homozygotes, as well as reductions in several direct and indirect binding partners of dysbindin-1 that contribute to the protein assembly known as BLOC-1. Studies of sdy mice on the original DBA/2J background and on a C57BL/6J background indicate that the $Dtnbp1^{sdy}$ mutation does not affect viability, basic sensory or motor functions, or measures of anxiety and motivation. Such studies do indicate, however, that the mutation affects several biological functions, including adrenal neurosecretion and pre- and postsynaptic aspects of dopaminergic, glutamatergic, and GABAergic transmission. These effects and those on prepulse inhibition, social interaction, and diverse aspects of spatial memory suggest that homozygous sdy mice may model various features of schizophrenia.

Keywords: BLOC-1; Dtnbp1; dysbindin; hippocampus; lysosome-related organelles; memory; prefrontal cortex; sandy mice; synapse; schizophrenia

Dysbindin-1 in schizophrenia

The dystrobrevin binding protein 1 (*DTNBP1*) gene encodes the first known paralog of the dysbindin protein family (dysbindin-1), which has three isoforms in humans (dysbindin-1A, -1B, and -1C) and two isoforms in mice (dysbindin-1A and -1C) (see Talbot et al., 2009). Shortly after the discovery of dysbindin-1 was reported (Benson et al., 2001), variation in *DTNBP1* was reported to be associated with schizophrenia (Straub et al., 2002). Such an association of this disorder with single nucleotide polymorphisms (SNPs) or combinations of them (haplotypes) in *DTNBP1* has now been reported in 17 other studies across the globe (see Talbot et al., 2009). It is not surprising, then, that *DTNBP1* is among the top candidate genes for schizophrenia (Allen et al., 2008; Sun et al., 2008).

While it is unclear if *DTNBP1* risk SNPs and haplotypes affect dysbindin-1 levels, there is growing evidence that dysbindin-1 levels are commonly reduced in at least three brain areas

*Corresponding author.
Tel.: +1 215 746 3647; Fax: +1 215 573 2041;
E-mail: talbotk2@mail.med.upenn.edu

DOI: 10.1016/S0079-6123(09)17910-4

dysfunctional in schizophrenia, namely the dorsolateral prefrontal cortex, the superior temporal gyrus, and the hippocampal formation (Talbot et al., 2004, 2009; Tang et al., 2009). As a result, there has been increasing interest in determining the functions of dysbindin-1 to understand how its reduction may contribute to clinical features of schizophrenia. Our ability to discover those functions in living animals is limited by the absence of known pharmacologic agents affecting the levels of dysbindin-1 or its protein–protein interactions and by the absence of dysbindin-1 knockout mice. Fortunately, however, we do have an animal model of dysbindin-1 function, namely the sandy or sdy mouse. We review here what we know about that animal as of July, 2009 and the clues it has offered to dysbindin-1 functions most relevant to schizophrenia.

The *Dtnbp1*^{sdy} mutation

At the Jackson Laboratory in 1983, a spontaneous autosomal recessive mutation affecting coat color occurred in DBA/2J mice. Unlike wild-type DBA/2J mice, which are brown (Fig. 1a), the homozygous mutants are sandy-colored (Fig. 1b). In the first study on sdy mice, Swank et al. (1991) established that the coat-color mutation occurs on chromosome 13 in the vicinity of cytogenic band A5. It was not until 2003, however, that the specific locus of the sdy mutation was published. In that year, Li et al. (2003) working in Swank's laboratory reported that the coat-color mutation occurs in the dystrobrevin binding protein 1 (mouse *Dtnbp1*) gene and is thus known as the *Dtnbp1*^{sdy} mutation. Transgenic addition of wild-type *Dtnbp1*to homozygous sdy mice essentially rescues the normal DBA/2J coat color and reverses another major feature of the mutant mice, namely platelet serotonin depletion associated with long bleeding time after tail snips (Li et al., 2003).

The *Dtnbp1*^{sdy} mutation is an in-frame deletion of 38,129 nucleotides from nucleotide number 3701 in intron 5 to nucleotide 12,377 in intron 7 that excises exons 6 and 7 (Fig. 1c, Li et al., 2003). This is a null protein mutation that results in

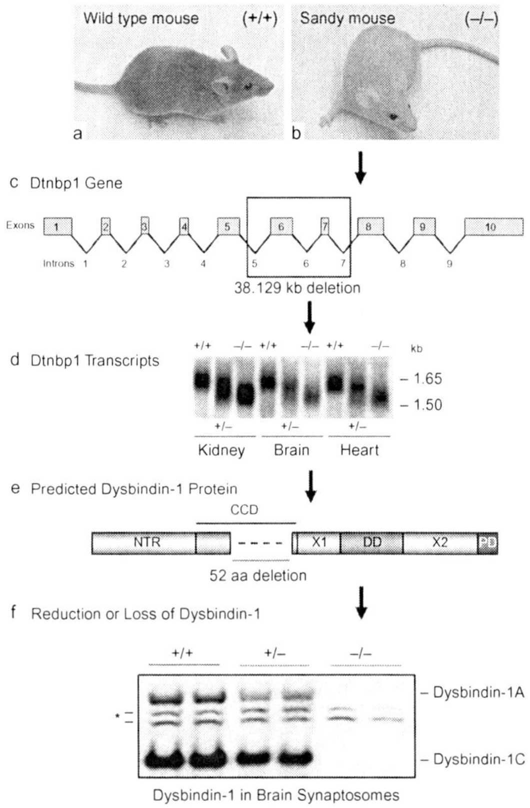

Fig. 1. The *Dtnbp1*^{sdy} mutation in DBA/2J mice and its effect on Dtnbp1 gene transcripts and tissue levels of dysbindin-1. (a) and (b) Coat-color difference between wild-type and sdy homozygous mice due to defective melanosome formation. (c) Basic exonic–intronic structure of mouse *Dtnbp1* gene showing large deletion mutation of exons 6 and 7 in sdy mice reported by Li et al. (2003). (d) Northern blots on the kidney, brain, and heart showing relative size of Dtnbp1 transcripts in wild-type (+/+), heterozygous (–/+), and homozygous (–/–) mice (Li et al., 2003). (e) Schematic of dysbindin-1 lacking 52 aa in the CCD due to mutation-induced truncation of transcript. CCD=coiled cool domain, DD=dysbindin domain, NTR=N-terminal region, PD=PEST domain, X1 and X2=uncharacterized zones. (f) Dysbindin-1A and -1C levels in brain synaptosomes of wild-type, heterozygous, and homozygous sdy mice seen in Western blot with antibody PA3111. Dysbindin-1B is not expressed in the mouse. Asterisk indicates bands with proteins other than dysbindin-1 (see Talbot et al., 2009). We thank Richard T. Swank for panels (a) and (b). Panels (c), (d), and (f) are adapted with permission from Li et al. (2003).

reduced levels of dysbindin-1A and -1C in heterozygotes and in undetectable levels of those proteins in homozygotes (Fig. 1f, Li et al., 2003; Talbot et al., 2009). *Dtnbp1*^{sdy} is not a null mRNA mutation, however, because the mutated gene is still transcribed. The transcripts are simply shorter (Fig. 1d) due to the absence of nucleotides encoding a 52 amino acid segment of the coiled coil domain (CCD) in wild-type dysbindin-1 (Fig. 1e). If translated, the mutant protein is predicted to have a CCD 58 amino acids shorter than the wild-type CCD, which may impair dysbindin-1's ability to form stable complexes with its normal binding partners (Li et al., 2003; Talbot et al., 2009). This is one of several possible reasons why the *Dtnbp1*^{sdy} mutation leads to loss of even a mutant dysbindin-1. The actual reasons are still unknown, but could also include aborted translation and protein misfolding resulting in targeted degradation.

Sdy/DBA and sdy/BL6 mice

As noted above, the original sdy mice were *Dtnbp1* mutants of the DBA/2J strain. This strain, unlike C57BL/6 mice, is homozygous for six mutations related to neurological, melanogenic, and/or inflammatory disorders as documented by Cox et al. (2009) and Talbot et al. (2009). These include cadherin 23^{ahl} (Cdh23^{ahl}=Cdh^{753A}) associated with an age-related hearing loss, as well as glycoprotein (transmembrane) nmb^{R150X} (Gpnmb^{R150X}) and tyrosinase-related protein 1^{isa} (Tyrp1^{isa}) associated with pigmentary glaucoma. These mutations must be present in sdy mice derived from DBA/2J animals (i.e., sdy/DBA mice), though not all such mice are necessarily homozygous for them since some have wild-type Gpnmb (Anderson et al., 2002).

These various mutations in DBA/2J mice appear to account for several developmental abnormalities in such animals compared to C57BL/6 mice as discussed elsewhere (Talbot et al., 2009). DBA/2J mice develop a high-frequency hearing loss between 3 and 4 weeks of age. By 5 weeks, hearing loss is detected at all frequencies tested and persists through old age. Between 3 and 4 months, at least some DBA/2J mice begin showing symptoms of glaucoma. As early as 6–7 months, DBA/2J mice have abnormal irises. Since sdy/DBA mice are necessarily homozygous for most of the alleles causing these conditions, they must share most of the auditory and visual deficits developing in DBA/2J animals. They are also expected to share the enhanced responses of DBA/2J mice to stress and decreased responses to dopamine agonists, as well as cognitive deficits on auditory, olfactory, and visual tasks of DBA/2J compared to C57BL/6 mice.

To avoid the confounding variables presented by the DBA/2J mutations affecting brain function, several groups have transferred the *Dtnbp1*^{sdy} mutation onto a C57BL/6J genetic background to create sdy/BL6 mice. These are not commercially available, but can be generated from sdy/DBA animals recovered from cryopreserved embryos (stock 001594 at the Jackson Laboratory [Bar Harbor, ME]). On both the DBA/2J and C57BL/6J backgrounds, sdy mice are fully viable and appear normal in many respects apart from depigmentation of hair and eyes. Both heterozygous (sdy/+) and homozygous (sdy/sdy) animals are normal in body weight, hair quality and density, number and length of whiskers, basic sensory abilities, neuromuscular strength, and sensorimotor reflexes (Hattori et al., 2008; Takao et al., 2008; Cox et al., 2009; Jentsch et al., 2009). In many other respects, they differ from littermate controls as described below.

Abnormalities of sdy mice compared to wild-type littermates

Except for two studies on sdy/BL6 mice completed thus far (Cox et al., 2009; Halene et al., 2009), abnormalities in sdy mice are known from studies on sdy/DBA animals. We summarize those abnormalities here.

Biogenesis of lysosome-related organelles complex 1 (BLOC-1) proteins

Dysbindin-1 is a component of BLOC-1, a protein assembly consisting of at least eight proteins that plays a critical role in the biogenesis of lysosome-related organelles (LROs) by promoting delivery of proteins those organelles require for functional maturation (Dell'Angelica, 2004; Setty et al., 2007). Among the many types of LROs are melanosomes, platelets, and probably endosomally derived synaptic vesicles (Talbot et al., 2009). Dysbindin-1 has many known and candidate binding partners, including four other BLOC-1 components: muted, snapin, pallidin, and BLOC-1 subunit 2 [BLOS2]). In the absence of any one BLOC-1 component, levels of the others are reduced as if they were stable only when part of the complex (Li et al., 2003). In sdy mice, then, there are reductions not only in dysbindin-1, but in other BLOC-1 proteins as shown in homozygous sdy/DBA mice, which have low levels of muted, pallidin, and snapin (cf., Li et al., 2003; Feng et al., 2008). It is thus important for future studies on sdy mice to test if observed abnormalities reflect dysbindin-1 reductions themselves as indicated by rescue in sdy mice with transgenic addition of wild-type *Dtnbp1* (see Li et al., 2003) and, if so, whether the abnormality is mediated by BLOC-1 as indicated by replication of the abnormality in other BLOC-1 mutants (e.g., muted mice [see Iizuka et al., 2007]).

LROs

Presumably due to a reduction in (or absence of) BLOC-1, sdy/DBA mice display abnormalities in several LROs, namely kidney secretory lysosomes, melanosomes, and both the dense granules and lysosomal granules of platelets (Swank et al., 1991). These abnormalities are discussed at length by Talbot et al. (2009). The abnormalities in melanosomes impair maturation of these pigment granules, accounting for the very light pigmentation of the hair and eyes of homozygous sdy mice. The abnormalities in platelet dense and lysosomal granules, which include depletion of serotonin stores, impair platelet aggregation, accounting for the long bleeding time of these mice following a tail snip.

Adrenal neurosecretion

Using electron microscopy and amperometry, Chen et al. (2008) studied the large dense core vesicles (LDCVs) of adrenal chromaffin cells in homozygous sdy/DBA mice. The sdy mice showed an increase in the size of these epinephrine- or norepinephrine-containing vesicles and in the amount of catecholamine molecules each released upon 80 mM KCl stimulus. But these mice also showed a lower density of LDCVs, slower kinetics of fusion pore closure, and a decreased frequency of stimulus-induced fusion events indicative of a decreased probability of release upon stimulation. This occurred in the apparent absence of any abnormalities in LDCV endocytosis and refilling, as well as in the absence of any alteration in levels of many proteins involved in exocytosis (complexin 1 or 2, munc18-1, SNAP25, synaptotagmin 1, syntaxin 1, or VAMP2). The abnormalities in LDCV size, number, catecholamine load, and fusion competence may reflect loss of BLOC-1 effects on biogenesis of reserve pool vesicles from which many readily releasable LDCVs may ultimately derive (see Talbot et al., 2009). The decreased probability of LDCV release may reflect loss of interaction between dysbindin-1 and its binding partner snapin, which normally boosts the number of LDCVs kept in a readily releasable state (Tian et al., 2005) and enhances efficient, synchronous release of synaptic vesicles (Pan et al., 2009).

Synaptic transmission

Dysbindin-1 is ubiquitous in neurons of the brain, but is enriched in certain neuronal populations and synaptic fields where dopamine, glutamate, and/or GABA act as neurotransmitters (Talbot et al., 2006, 2009). In synaptosomal fractions of mouse and human brains, dysbindin-1A is virtually restricted to postsynaptic densities (PSDs), dysbindin-1B (not present in mice) is almost entirely associated with synaptic vesicles, and dysbindin-1C is present to some extent with synaptic vesicles and to a larger extent with PSDs (Talbot et al., 2009).

Dopamine release is normally suppressed by dysbindin-1 according to in vitro studies on PC-12 cells (Kumamoto et al., 2006). It is not surprising, then, that there is evidence of increased dopamine release in homozygous sdy/DBA mice, specifically in tissue-containing limbic targets of midbrain dopamine neurons (Murotani et al., 2007). Especially since loss of dysbindin-1 can cause overexpression of cell surface D2 dopamine receptors (D2Rs, Iizuka et al., 2007), such mice may model the dopaminergic hypothesis of schizophrenia according to which its positive symptoms are associated with hyperactivity of the mesolimbic dopaminergic pathway and with hyperstimulation of D2Rs.

As in the case of dopamine transmission, dysbindin-1 has both pre- and postsynaptic effects on glutamatergic transmission. But unlike the presynaptic inhibition of dopamine release, glutamate release is facilitated by dysbindin-1 according to in vitro studies on primary cortical neurons (Numakawa et al., 2004). Consistent with those studies, glutamatergic transmission is impaired in homozygous sdy/DBA mice. As in adrenal LDCVs of such mice, Chen et al. (2008) found that synaptic vesicles in glutamatergic synapses of hippocampal field CA1 in homozygous sdy/DBA mice were larger and held more transmitter, but were less numerous in the reserve pool and appeared to have slower release kinetics and decreased probability of release. This last effect might be due to loss of dysbindin-1 interactions with snapin and to reported snapin reductions in the hippocampus of these mice (Feng et al., 2008). Impaired glutamate release would help explain why Chen et al. (2008) found that CA1 pyramidal cells in homozygous sdy/DBA mice were less affected by stimulation of their Schaffer collateral input as shown by smaller evoked excitatory postsynaptic currents (eEPSCs), which may involve not only presynaptic, but also postsynaptic abnormalities given the greater thickness of PSDs found in these mice by the same investigators. Reduced activation of CA1 pyramidal cells with subsequently reduced activation of interneurons innervated by those cells may contribute to the marked absence of inhibitory responses in CA1 of homozygous sdy/DBA and sdy/BL6 mice to excitation by Schaffer collateral stimulation (G. Carlson et al., in press).

Similar, but not identical, abnormalities in sdy/DBA mice are reported by Jentsch et al. (2009) in deep pyramidal cells of the prelimbic cortex, the putative rodent homolog of the human dorsolateral prefrontal cortex. As in CA1, they found evidence of impaired synaptic glutamate release by pyramidal cells evident in reduced paired-pulse facilitation in the homozygous mice. Such an impairment in the pyramidal cells was consistent with the fact that they displayed (1) less frequent miniature EPSCs (mEPSCs) in both heterozygous and homozygous mice and (2) smaller amplitude mEPSCs and eEPSCs in homozygous mice. The EPSC alterations indicate abnormally low basal levels of excitatory input on the deep pyramidal cells despite their lower spike thresholds in both the heterozygote and homozygote mice. This may be due to the reduced GABAergic input observed in the prelimbic homozygous sdy mice (Trantham-Davidson et al., 2008). Such input derives in part from fast-spiking, parvalbumin-containing interneurons, which drive gamma oscillations in the EEG via their innervation of pyramidal neuron cell bodies and axon initial segments. In the prelimbic cortex of homozygous sdy/DBA mice, these interneurons are less excitable (Ji et al., 2008) and display low basal levels of excitation (Trantham-Davidson et al., 2008). Since the excitability of these interneurons is driven by NMDA receptor channels, sdy mice may model the NMDA receptor hypofunction of fast-spiking interneurons increasingly considered a feature of schizophrenia and a cause of impaired induction of gamma oscillations common in that disorder.

Behavior

As noted earlier, sdy/DBA and sdy/BL6 mice are normal in basic sensory and motor functions, though the former display elevated sensitivity to thermally induced pain (Bhardwaj et al., 2009) and the latter display reduced auditory prepulse inhibition (PPI, Halene et al., 2009). Neither the sdy/DBA nor the sdy/BL6 mice display consistently observed evidence of anxiety in the elevated plus maze or the light/dark box transition test (Hattori et al., 2008; Takao et al., 2008; Cox et al., 2009). As summarized in Table 1, however, the behavior of homozygous sdy mice is abnormal in activity, social interactions, and diverse tests of memory. Abnormal activity is often reported, but both hypoactivity (Hattor et al., 2008; Takao et al., 2008) and hyperactivity due to decreased habituation to the open field

Table 1. Behavioral abnormalities detected in sandy mice compared to wild-type littermates[a]

Behavioral test	DBA/2J background		C57BL/6J background		References
	sdy/+	sdy/sdy	sdy/+	sdy/sdy	
Elevated plus maze					
Time spent in open arms	–	ND	–	–	Hattori et al. (2008)
	–	–	ND	ND	Cox et al. (2009)
Entries into open arms	–	↓	–	–	Hattori et al. (2008)
	–	ND	–	–	Takao et al. (2008)
Distance traveled in open arms	–	ND	–	–	Hattori et al. (2008)
	–	ND	–	–	Takao et al. (2008)
	–	–	↑	↑	Cox et al. (2009)
Light/dark transition	–	ND[b]	–	–	Takao et al. (2008)
Locomotion in open field	–	ND	–	–	Feng et al. (2008)
	–	↓	–	–	Hattori et al. (2008)
	–	↓	–	–	Takao et al. (2008)
	ND	↑	–	–	Bhardwaj et al. (2009)
	–	–	ND	↑	Cox et al. (2009)
Locomotor response to amph[c]					
Single dose	↓	↓	–	–	Bhardwaj et al. (2009)
Repeated doses	ND	↑	–	–	Bhardwaj et al. (2009)
Prepulse inhibition	ND	ND	–	–	Li et al. (2003)
	–	–	ND	↓	Halene et al. (2009)
Social interactions					
Number of contacts	–	↓	–	–	Hattori et al. (2008)
Duration of contacts	–	↓	–	–	Feng et al. (2008); Hattori et al. (2008)
Thermal pain sensitivity	↑	↑	–	–	Bhardwaj et al. (2009)
Contextual fear memory	ND	↑	–	–	Bhardwaj et al. (2009)
Novel object recognition memory	↓	↓	–	–	Feng et al. (2008) Bhardwaj et al. (2009)
Spatial memory					
Barnes circular maze	–	↓[d]	–	–	Takao et al. (2008)
Morris water maze	–	↓	–	–	Jentsch et al. (2007)
	–	–	ND	↓	Cox et al. (2009)
Non-match to position	ND	↓	–	–	Jentsch et al. (2009)
T-maze forced alternation	–	↓	–	–	Takao et al. (2008)

[a]Dash (–) indicates animal genotype not tested; ND=no difference from wild-type littermate.
[b]Sdy were normal except that they traveled more in the dark box.
[c]Amph=amphetamine (2.5 mg/kg ip).
[d]Effect noted with 7-day (not 24-h) delay between training and test.

(Bhardwaj et al., 2009; Cox et al., 2009) have been observed. Social interactions are reduced in number and duration (Hattori et al., 2008; Feng et al., 2008) that may reflect social withdrawal (a negative symptom of schizophrenia). While they display no evidence of reduced motivation and only limited deficits in learning itself (Takao et al., 2008; Bhardwaj et al., 2009; Cox et al., 2009; Jentsch et al., 2009), sdy mice do display evidence of altered memory capacities. They show heightened memory of signals associated with delivery of painful stimuli (Bhardwaj et al., 2009) but clear deficits in novel object recognition memory (Feng et al., 2008; Bhardwaj et al., 2009) and in diverse types of spatial memory (Jentsch et al., 2007, 2009; Takao et al., 2008; Cox et al., 2009). These deficits may be due in part to impaired generation of long-term potential in homozygous sdy/DBA (Glen et al., 2008) and sdy/BL6 (G. Carlson et al., in press).

Does the sdy mouse model aspects of schizophrenia?

While the *Dtnbp1*^{sdy} mutation has not been found in schizophrenia cases, homozygous sdy mice share multiple biological, behavioral, and cognitive features of schizophrenia. Like homozygous sdy mice as described above, schizophrenia cases display reduced synaptic dysbindin-1 (Talbot et al., 2004, 2009; Tang et al., 2009), evidence of hyperactive mesolimbic dopamine pathways (Laruelle, 2003), and disrupted glutamatergic–GABAergic interactions in prefrontal cortex (Lewis and Moghaddam, 2006). Schizophrenia cases also display the reduced PPI, decreased social interactions, and multiple cognitive deficits of homozygous sdy mice described above that are shared with other proposed mouse models of schizophrenia (Powell and Miyakawa, 2006; Mazzoncini et al., 2009). Indeed, the spatial memory deficits displayed by homozygous mice in the Morris water maze (Cox et al., 2009) and the T-maze (Takao et al., 2008) resemble those displayed by schizophrenia cases in a virtual Morris water maze task (Hanlon et al., 2006) and in other tests of spatial working memory (Glahn et al., 2003). Consequently, homozygous sdy mice potentially model diverse aspects of schizophrenia.

Acknowledgments

Work on this review was supported by US National Institutes of Health grants (MH072880 and MH064045).

References

Allen, N. C., Bagade, S., McQueen, M. B., Ioannidis, J. P. A., Kavvoura, F. K., Khoury, M. J., et al. (2008). Systematic meta-analyses and field synopsis of genetic association studies in schizophrenia: The SzGene database. *Nature Genetics*, *40*, 827–834.

Anderson, M. G., Smith, R. S., Hawes, N. L., Zabaleta, A., Chang, B., Wiggs, J. L., et al. (2002). Mutations in genes encoding melanosomal proteins cause pigmentary glaucoma in DBA/2J mice. *Nature Genetics*, *30*, 81–85.

Benson, M. A., Newey, S. E., Martin-Rendon, E., Hawkes, R., & Blake, D. J. (2001). Dysbindin, a novel coiled-coil-containing protein that interacts with the dystrobrevins in muscle and brain. *Journal of Biological Chemistry*, *276*, 24232–24241.

Bhardwaj, S. K., Baharnoori, M., Sharif-Askari, B., Kamath, A, Williams, S., & Srivastava, L. K. (2009). Behavioral characterization of dysbindin-1 deficient sandy mice. *Behavioural Brain Research*, *197*, 435–441.

Chen, X.-W., Feng, Y.-Q., Hoo, C.-J., Guo, X.-L., He, X., Zhou, Z.-Y., et al. (2008). DTNBP1, a schizophrenia susceptibility gene, affects kinetics of transmitter release. *Journal of Cell Biology*, *181*, 791–801.

Cox, M. M., Tucker, A. M., Tang, J., Talbot, K., Richer, D. C., Yeh, L., et al. (2009). Neurobehavioral abnormalities in the sandy mouse, a dysbindin-1 mutant, on a C57BL/6J background. *Genes, Brain and Behavior*, *8*, 390–397.

Dell'Angelica, E. C. (2004). The building BLOC(k)s of lysosomes and related organelles. *Current Opinion in Cell Biology*, *16*, 458–464.

Feng, Y.-Q., Zhou, Z.-Y., He, X., Wang, H., Guo, X.-L., Hao, C.-J., et al. (2008). Dysbindin deficiency in sandy mice causes reduction of snapin and displays behaviors related to schizophrenia. *Schizophrenia Research*, *106*, 218–228.

Glahn, D. C., Therman, S., Manninen, M., Huttunen, M., Kaprio, J., Lönnqvist, J., et al. (2003). Spatial working memory as an endophenotype for schizophrenia. *Biological Psychiatry*, *53*, 624–626.

Halene, T.B., Amann, L.C., Ehrlichman, R.S., Lin, R., Kazi, H., Talbot, K., et al. (2009) Neurobehavioral abnormalities in dysbindin-1 mutant mice on a C57BL/6J background. (Presentation at 2009 Society for Neuroscience Annual Meeting, Chicago, IL)

Hanlon, F. M., Weisend, M. P., Hamilton, D. A., Jones, A. P., Thoma, R. J., Huang, M., et al. (2006). Impairment on the hippocampal-dependent virtual Morris water task in schizophrenia. *Schizophrenia Research*, *87*, 67–80.

Hattori, S., Murotani, T., Matsuzaki, S., Ishizuka, T., Kumamoto, N., Takeda, M., et al. (2008). Behavioral abnormalities and dopamine reductions in sdy mutant mice with a deletion in *Dtnbp1*, a susceptibility gene for schizophrenia. *Biochemical and Biophysical Research Communications*, *373*, 298–302.

Iizuka, Y., Sei, Y., Weinberger, D. R., & Straub, R. E. (2007). Evidence that the BLOC-1 protein dysbindin modulates dopamine D2 receptor internalization and signaling but not D1 internalization. *Journal of Neuroscience*, *27*, 12390–12395.

Jentsch, J.D., Tinsley, M., Jairl, C., Horowitz, B., Seu, E., & Cannon, T. (2007) Null mutation of the gene coding for dysbindin is associated with poor working memory and spatial learning in mice. (Abstract 59.16 from 2007 Society for Neuroscience Annual Meeting in San Diego, CA)

Jentsch, J. D., Trantham-Davidson, H., Jairl, C., Tinsley, M., Cannon, T. D., & Lavin, A. (2009). Dysbindin modulates prefrontal cortical glutamatergic circuits and working memory function in mice. *Neuropsychopharmacology*, *34*, 2601–2608.

Ji, Y., Yang, F., Gao, W., & Lu, B. (2008) Alterations in dopamine D2 receptor trafficking and D2 regulation of cortical neurons in sandy mice. (Abstract 254.23 from *2008* Society for Neuroscience Annual Meeting, Washington, DC.)

Kumamoto, N., Matsuzaki, S., Inoue, K., Hattori, T., Shimizu, S., Hashimoto, R., et al. (2006). Hyperactivation of midbrain dopaminergic system in schizophrenia could be attributed to the down-regulation of dysbindin. *Biochemical and Biophysical Research Communications*, *345*, 904–909.

Laruelle, M. (2003). Dopamine transmission in the schizophrenic brain. In S. R. Hirsch & D. R. Weinberger (Eds.), *Schizophrenia* (pp. 365–387). Malden, MA: Blackwell Science Ltd..

Lewis, D. A., & Moghaddam, B. (2006). Cognitive dysfunction in schizophrenia, convergence of γ-aminobutyric acid and glutamate alterations. *Archives of Neurology*, *63*, 1372–1376.

Li, W., Zhang, Q., Oiso, N., Novak, E. K., Gautam, R., O'Brien, E. P., et al. (2003). Hermansky-Pudlak syndrome type 7 (HPS-7) results from mutant dysbindin, a member of the biogenesis of lysosome-related organelles complex 1 (BLOC-1). *Nature Genetics*, *35*, 84–89.

Mazzoncini, R., Zoli, M., Tosato, S., Lasalvia, A., & Ruggeri, M. (2009). Can the role of genetic factors in schizophrenia be enlightened by studies of candidate gene mutant mice behavior? *World Journal of Biological Psychiatry*, *10* (in press).

Murotani, T., Ishizuka, T., Hattori, S., Hashimoto, R., Matsuzaki, S., & Yamatodani, A. (2007). High dopamine turnover in the brains of Sandy mice. *Neuroscience Letters*, *421*, 47–51.

Numakawa, T., Yagasaki, Y., Ishimoto, T., Okada, T., Suzuki, T., Iwata, N., et al. (2004). Evidence of novel neuronal functions of dysbindin, a susceptibility gene for schizophrenia. *Human Molecular Genetics*, *13*, 2699–2708.

Pan, P.-Y., Tian, J.-H., & Sheng, Z.-H. (2009). Snapin facilitates the synchronization of synaptic vesicle fusion. *Neuron*, *61*, 412–424.

Powell, C. M., & Miyakawa, T. (2006). Schizophrenia-relevant behavioral testing in rodent models: A uniquely human disorder? *Biological Psychiatry*, *59*, 1198–1207.

Setty, S. R. G., Tenza, D., Truschel, S. T., Chou, E., Sviderskaya, E. V., Theos, A. C., et al. (2007). BLOC-1 is required for cargo-specific sorting from vacuolar early endosomes toward lysosome-related organelles. *Molecular Biology of the Cell*, *18*, 768–780.

Straub, R. E., Jiang, Y., MacLean, C. J., Ma, Y., Webb, B. T., Myakishev, M. V., et al. (2002). Genetic variation in the 6p22.3 gene DTNBP1, the human ortholog of the mouse dysbindin gene, is associated with schizophrenia. *American Journal of Human Genetics*, *71*, 337–348.

Sun, J., Kuo, P.-H., Riley, B. P., Kendler, K. S., & Zhao, Z. (2008). Candidate genes for schizophrenia: A survey of association studies and gene ranking. *American Journal of Medical Genetics B (Neuropsychiatric Genetics)*, *147B*, 1173–1181.

Swank, R. T., Sweet, H. O., Davisson, M. T., Reddington, M., & Novak, E. K. (1991). Sandy: A new mouse model for platelet storage pool deficiency. *Genetics Research*, *58*, 51–62.

Takao, K., Toyama, K., Nakanishi, K., Hattori, S., Takamura, H., Takeda, M., et al. (2008). Impaired long-term memory retention and working memory in sdy mutant mice with a deletion in Dtnbp1, a susceptibility gene for schizophrenia. *Molecular Brain*, *1*, 11.

Talbot, K., Cho, D.-S., Ong, W.-Y., Benson, M. A., Han, L.-Y., Kazi, H. A., et al. (2006). Dysbindin-1 is a synaptic and microtubular protein that binds brain snapin. *Human Molecular Genetics*, *15*, 3041–3054.

Talbot, K., Eidem, W. L., Tinsley, C. L., Benson, M. A., Thompson, E. W., Smith, R. J., et al. (2004). Dysbindin is reduced in intrinsic, glutamatergic terminals of the hippocampal formation in schizophrenia. *Journal of Clinical Investigation*, *113*, 1353–1363.

Talbot, K., Ong, W. Y., Blake, D. J., Tang, J., Louneva, N., Carlson, G. C., et al. (2009). Dysbindin-1 and its protein family with special attention to the potential role of dysbindin-1 in neuronal functions and the pathophysiology of schizophrenia. In D. Javitt & J. Kantorowitz (Eds.), *Handbook of Neurochemistry and Molecular Neurobiology* (3rd ed., Vol. 27, US, New York: Springer, pp. 107–241.

Tang, J., LeGros, R. P., Louneva, N., Yeh, L., Cohen, J. W., Hahn, C.-G., et al. (2009). Dysbindin-1 in dorsolateral prefrontal cortex of schizophrenia cases is reduced in an isoform-specific manner unrelated to dysbindin-1 mRNA expression. *Human Molecular Genetics*, *18*, 3851–3863.

Tian, J.-H., Wu, Z.-X., Unzicker, M., Lu, L., Cai, Q., Li, C., et al. (2005). The role of snapin in neurosecretion: *Snapin* knock-out mice exhibit impaired calcium-dependent exocytosis of large dense-core vesicles in chromaffin cells. *Journal of Neuroscience*, *25*, 10546–10555.

Trantham-Davidson, H., Jentsch, J.D., and Lavin, A. (2008) Effects of null mutation of the gene encoding dysbindin-1 on cortical fast-spiking interneurons (Abstract 254.8 from 2008 Society for Neuroscience Annual Meeting, Washington, DC.)

SECTION IV

Other than Rodents

A. Sawa (Ed.)
Progress in Brain Research, Vol. 179
ISSN 0079-6123

CHAPTER 11

Zebrafish: a model system to examine the neurodevelopmental basis of schizophrenia

Jill A. Morris*

Program in Human Molecular Genetics, Department of Pediatrics, Feinberg School of Medicine, Children's Memorial Research Center, Northwestern University, Chicago, IL, USA

Abstract: Schizophrenia is a devastating disorder caused by both genetic and environmental factors that disrupt brain development and function. It is distinguished as a neurodevelopmental disorder in part due to early cognitive impairments, behavioral dysfunction in childhood and adolescence, and abnormalities in central nervous system development. Zebrafish are recognized as an important vertebrate model for human development and disease. There are many advantages of using zebrafish as a model, such as low cost to maintain, rapid life cycle, optical clarity and rapid external embryonic development. Furthermore, multiple molecular genetic techniques have been developed to readily study gene function during development. In this review, we will discuss the advantages of using the zebrafish model system to study schizophrenia.

Keywords: schizophrenia; zebrafish; neurodevelopment; neurogenesis; neuronal migration; cell fate; Disc1; Disrupted-In-Schizophrenia 1; Neuregulin 1

Introduction

In the 1970s, Dr. George Streisinger at the University of Oregon worked diligently to establish zebrafish (*Danio rerio*), a teleost fish, as a vertebrate model to study the development of the nervous system [reviewed in (Grunwald and Eisen, 2002)]. His research along with that of his colleagues has resulted in the establishment of zebrafish as an important vertebrate model to study human disease including heart disease (Chico et al., 2008), cancer (Amatruda et al., 2002; Amatruda and Patton, 2008), motor neuron disease (Beattie et al., 2007) and Alzheimer's disease (Newman et al., 2007). Furthermore, it is being established as a tool for high-throughput toxicology and drug discovery screens (Kari et al., 2007).

There are many advantages of using zebrafish as a model including low cost to maintain, rapid life cycle, large number of progeny and external transparent embryonic development. A zebrafish mating pair can produce several hundred progeny in a single mating. These progeny produced externally will undergo rapid embryonic development from egg to self-feeding, swimming larvae in five days. These features make them an ideal vertebrate model for genetic and behavioral screens. In addition, their transparency during

*Corresponding author.
Tel.: +1 773 755 6351; Fax: +1 773 755 6345;
E-mail: j-morris4@northwestern.edu

DOI: 10.1016/S0079-6123(09)17911-6

embryonic development allows for the characterization of developmental defects as well as the study of cell behaviors using fluorescent proteins. Furthermore, multiple molecular techniques have been developed to examine gene function during development.

This review will focus on the use of zebrafish to study schizophrenia. Schizophrenia is a debilitating neurodevelopmental illness, which affects 1% of the population and is associated with high rates of morbidity and mortality (Lewis and Lieberman, 2000; Harvey et al., 2002). It is distinguished as a neurodevelopmental disorder as patients demonstrate cognitive and behavioral dysfunction in childhood and adolescence, abnormalities in central nervous system (CNS) development and no demonstrative neurodegeneration (Marenco and Weinberger, 2000; Lewis and Levitt, 2002; Sawa and Snyder, 2002). The genetic contribution to schizophrenia has been established through twin and family studies (Gottesman and Erlenmeyer-Kimling, 2001). Although there is an environmental component to the risk of developing schizophrenia, it is highly dependent on genetics (Sullivan et al., 2003). Therefore, determining the genetic and developmental basis of schizophrenia is critical for understanding disease pathogenesis and identifying new treatments. The zebrafish model system allows for the study of both the genetic and developmental basis of disease. This review will discuss the benefits of using zebrafish as a model system with emphasis on advantages specific to studying the pathogenesis of schizophrenia and recent research using zebrafish to study the function of schizophrenia susceptibility genes.

Zebrafish genome

The zebrafish genome contains approximately 2 gigabases on 25 chromosomes. The Sanger Institute is currently sequencing the zebrafish genome with the latest release containing over 14,000 annotated Vega genes (http://www.sanger.ac.uk/Projects/D_rerio/). Due to a whole genome duplication event in teleost, there is a subset of zebrafish genes that are duplicated resulting in two orthologs of a human gene (Taylor et al., 2003; Woods et al., 2005). Furthermore, these paralogs may have different expression patterns and divided or novel functions. For example, there are two orthologs of the human NudE-like gene (NDEL1/NUDEL) in zebrafish (Drerup et al., 2007). NDEL1 is the mammalian homolog of nuclear distribution molecule, NudE, which functions in nuclear migration during hyphal growth in *Aspergillus nidulans* (Efimov and Morris, 2000). In mammals, NDEL1 functions in nucleokinesis, neuronal migration and cortical development [reviewed in (Wynshaw-Boris, 2007)]. It has also been demonstrated that NDEL1 interacts with DISC1 (Disrupted-In-Schizophrenia 1), a schizophrenia susceptibility gene (Millar et al., 2003; Morris et al., 2003; Ozeki et al., 2003; Brandon et al., 2004; Kamiya et al., 2006). The zebrafish orthologs of *NDEL1*, *ndel1a* and *ndel1b*, have non-overlapping expression patterns during embryonic development (Drerup et al., 2007). In situ hybridization analysis reveals that *ndel1a* is expressed in the anterior CNS, trigeminal ganglia and eyes during embryonic development. In contrast, *ndel1b* is first expressed in the developing somites and then later in the developing brain. These spatial and temporal differences in gene expression may suggest functional difference between these paralogs.

Molecular genetic techniques in zebrafish

Multiple schizophrenia susceptibility genes have been identified including *neuregulin 1* (Stefansson et al., 2002, 2003), *COMT* (*catechol O-methyl transferase*) (Egan et al., 2001; Bilder et al., 2002), *dysbindin* (Straub et al., 2002) and DISC1 (*Disrupted In Schizophrenia 1*) (Millar et al., 2000). There are established molecular genetic techniques available to readily study the function of these susceptibility genes in zebrafish. Morpholino oligonucleotides (MOs) are used in zebrafish to transiently decrease the expression of a particular gene [reviewed in (Bill et al., 2009)]. MOs are designed anti-sense to a targeted RNA and contain 25 morpholine bases with a neutrally charged phosphorodiamidate backbone.

This results in a modified oligonucleotide with high binding affinity to the targeted RNA. MOs decrease gene expression by binding to the targeted transcript and interfering with RNA processing or translation. There are two types of MOs that can be designed to a transcript: translational blocking and splice blocking. Translational-blocking MOs are designed to 5′UTR (untranslated region) or start site of the targeted gene. They function by interfering with the translation apparatus through steric hindrance, subsequently knocking down endogenous protein levels (Nasevicius and Ekker, 2000). To determine the efficiency of a translational-blocking MO, an antibody to the protein of interest is used to demonstrate decrease in the endogenous protein expression. Splice-blocking MOs are designed to splice junctions and result in altered RNA processing (Draper et al., 2001; Knight et al., 2003). The effects of the pre-mRNA splicing can be characterized and quantified by RT-PCR. For example, to study the function of Disc1 in zebrafish, we designed an MO (MOE3) to the intron 2/exon 3 splice-acceptor site of the zebrafish *disc1* gene (Fig. 1A) (Drerup et al., 2009). MOE3 was injected into one-cell stage embryos and total RNA was isolated at multiple timepoints. We determined by RT-PCR that injection of the MOE3 resulted in a variant *disc1* transcript (Fig. 1B). We sequenced this product and confirmed that exon 3 is deleted producing a premature stop site.

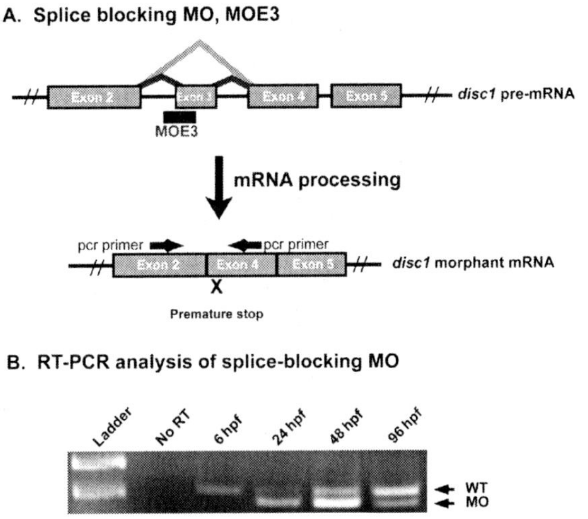

Fig. 1. The Disc1 MOE3 splice-blocking MO generates an altered *disc1* mRNA transcript. (A) A splice site MO was designed to the intron 2/exon 3 splice-acceptor site. Binding of the MO alters pre-mRNA processing, i.e. the removal of exon 3. A frameshift results in a premature stop site. Grey lines indicate altered splicing and black lines indicate normal splicing. (B) RT-PCR analysis demonstrates that injection of MOE3 into zebrafish embryo results in alternative splicing of the *disc1* transcript at 24, 48 and 96 hpf (hours post-fertilization). At 6 hpf, maternally derived *disc1* mRNA is present and is unaffected by MOE3. The figure is modified from Drerup et al. (2009).

MOs are usually introduced into the yolk of a zebrafish embryo at the 1–8 cell stage. As MOs transiently decrease the expression of a target gene, the resultant MO phenotypes are typically characterized within 3–5 dpf (days post-fertilization). MOs are not subject to enzymatic degradation, but their cellular levels are thought to be decreased through cell division during development. Off-target effects including developmental delay and p53-dependent cell death often result when using MOs (Robu et al., 2007). Therefore, off-target effects need to be carefully controlled for when using MOs. This can be done by (1) establishing a dose–response curve of the MO, (2) stage-matching control embryos, (3) demonstrating a second MO results in the same phenotype, (4) demonstrating that injection of suboptimal doses of two MOs results in an additive effect, (5) using a mismatch control MO and (6) demonstrating that the phenotype can be suppressed or rescued by co-injection of wild-type mRNA of your target gene. In addition, a tP53-targeted MO can be co-injected with the gene-specific MO to suppress the nonspecific p53 cell death due to MO toxicity (Robu et al., 2007).

In order to establish stable zebrafish lines with an interrupted or mutated endogenous gene, random mutagenesis by chemicals, retroviral insertion or radiation followed by genetic screens have been used in zebrafish [reviewed in (Amsterdam and Hopkins, 2006)]. Homologous recombination techniques in embryonic stem cells as employed in the generation of knockout mouse models are not established for zebrafish. Fortunately, targeted mutagenesis in the zebrafish has recently been established using zinc-finger nucleases (ZFNs) (Doyon et al., 2008; Meng et al., 2008). ZFNs are chimeric proteins that have both a DNA-binding zinc-finger domain and the *FokI* restriction endonuclease. By altering the specificity of the zinc fingers, a ZFN can be designed to target a double-stranded break in a specific endogenous gene. The use of ZFNs in zebrafish will allow for the determination gene function through loss of function in a stable fish line.

In addition to loss-of-function studies, methods have been developed to perform gain-of-function studies in zebrafish. Specifically, in vitro transcribed mRNA of a particular gene can be injected into the embryo, resulting in protein overexpression. The length of transient expression of the injected mRNA is dependent on the mRNA stability. Finally, due to short generation time, transgenic zebrafish can be rapidly produced. Multiple types of transgenic strategies have been implemented in zebrafish including fluorescent reporter lines, chemically inducible promoters such as the Tet-On system, heat-shock-induced promoters, the cre-lox system and the GAL4-UAS system [reviewed in (Deiters and Yoder, 2006)]. In addition, the use of transposons, mobile DNA elements that are used to aid integration into a genome, has greatly increased the rate of transgenesis in zebrafish. The Sleeping Beauty and Tol2 transposons increase the rate of transgenesis 30–50%, respectively (Davidson et al., 2003; Kawakami et al., 2004). All of these techniques will be valuable for the study of schizophrenia susceptibility genes during development.

Zebrafish as a model to study brain development

The overall structure of the teleost brain is similar to that of the mammalian brain. The teleost brain is divided into the forebrain including the diencephalon and telencephalon, midbrain and hindbrain. However, there are differences including an everted telencephalon and smaller cerebral hemispheres in zebrafish [reviewed in (Wullimann and Mueller, 2004)]. Teleosts have the major sensory systems including olfaction, vision, taste, touch, balance and hearing (Tropepe and Sive, 2003). They also have the major neurotransmitters systems including noradrenergic, dopaminergic and cholinergic (Wullimann and Rink, 2002; Rink and Wullimann, 2004; Wullimann and Mueller, 2004). In addition, teleosts have a similar structure to mammals for encoding spatial information (Rodriguez et al., 2002a, b). The lateral pallium in teleosts is similar to the hippocampus in mammals. Lesions to the lateral pallium in goldfish result in spatial memory deficits similar to lesions in the hippocampus of mammals (Rodriguez et al., 2002b).

The zebrafish model system can be used to study a multitude of neurodevelopmental processes that may be disrupted in schizophrenia pathogenesis including neurogenesis, neuronal migration and cell fate determination. Neurogenesis occurs in both the larval and adult zebrafish (Tropepe and Sive, 2003; Grandel et al., 2006; Zupanc, 2008; Lam et al., 2009). Adult neurogenesis in the teleost brain including regions homologous to the hippocampus and olfactory bulb in mammals has been extensively studied and I refer readers to multiple reviews describing this work in detail (Grandel et al., 2006; Zupanc, 2008; Lam et al., 2009).

One of the major advantages of zebrafish is the ability to monitor cell behavior such as neuronal migration in a live embryo using confocal microscopy. (Mione et al., 2008; Abraham et al., 2009). There are numerous zebrafish reporter lines available in which cell-type-specific promoters are driving the expression of fluorescent proteins. These transgenic lines allow for the examination of neuronal migration in a live embryo using time-lapse microscopy. Using these reporter lines, Mione et al. (2008) examined the migration of GABAergic interneurons and glutamatergic septal neurons of the telencephalon, mitral cell precursors from the dorsocaudal telencephalon to the olfactory bulb, and projection neurons and Purkinje cells of the cerebellum.

There are also multiple methods available to perform fate mapping in zebrafish. One potential mechanism to determine cell fate in the developing brain is to use a Kaede transgenic zebrafish.

Kaede is a fluorescent protein isolated from the stony coral *Trachyphylia geoffroyi*. This protein normally emits a green fluorescent signal, but upon UV (ultraviolet) irradiation the emission converts to a red fluorescence. In transgenic zebrafish lines expressing Kaede, one can photo-convert a small group of cells from green to red fluorescent using confocal microscopy and track the cells through developmental processes (Hatta et al., 2006). Another method is by photoactivating a fluorescent tracer using two-photon microscopy. Russek-Blum et al. (2009) determined a high-resolution fate map of the diencephalon in zebrafish using a cell-lineage tracer dye, DMNB-caged fluorescein.

Zebrafish as a model to study behavior

Schizophrenia is characterized by a multitude of positive and negative symptoms including hallucinations, delusions and social withdrawal as well as cognitive deficits (Lewis and Lieberman, 2000; Harvey et al., 2002). In the field of schizophrenia research, the mouse is typically used as a model to study the behavioral abnormalities seen in schizophrenia including deficits in working memory, impaired sensory motor gating and increased activity/hyperlocomotion (Lipska and Weinberger, 2000; Arguello and Gogos, 2006; Powell et al., 2009). It is difficult to generate an animal model that can recapitulate all of clinical symptoms of a complex behavioral disease like schizophrenia. However, zebrafish provide the unique opportunity to perform large forward genetic screens to identify new genes involved in a particular behavior in addition to studying behavior in a genetically characterized mutant line.

Forward genetic screens to identify genes involved in behavior have been performed in both larval and adult zebrafish (Burgess and Granato, 2008). In the adult, behavioral assays have been established for learning and memory (Darland and Dowling, 2001), conditioned place preference (Darland and Dowling, 2001), fear and anxiety (Gerlai et al., 2009), aggression (Gerlai, 2003) and social interactions (Darrow and Harris, 2004; Engeszer et al., 2004; Miller and Gerlai, 2007). Behaviors in larvae have also been examined including acoustic startle (Kimmel et al., 1974), escape response (Eaton et al., 1977), olfactory responses (Vitebsky et al., 2005), visually mediated behaviors (Fleisch and Neuhauss, 2006) and adaptation to environment (Burgess and Granato, 2007a).

Of particular interest to the field of schizophrenia are behavioral assays relating to learning and memory, locomotion and impaired sensory motor gating. Recently, Drs Burgess and Granato at the University of Pennsylvania established a high-throughput assay to measure prepulse inhibition (PPI) in zebrafish larvae (Burgess and Granato, 2007b). PPI, a type of sensorimotor gating, is the attenuation of a startle response by a preceding weaker non-startling stimulus. PPI is often measured in mouse models of schizophrenia to establish the relevance of a model to schizophrenia, examine gene function and test potential therapeutics (Powell et al., 2009). Zebrafish larvae have a strong startle response that is characterized with a "C-bend" of the body, a smaller counter bend, and then swimming (Kimmel et al., 1974). Drs Burgess and Granato developed a video tracking software that allows them to record and quantify the PPI in zebrafish larvae (Burgess and Granato, 2007b). The software allows them to record 2500 responses per hour of 30 larvae. They demonstrated that the acoustic startle response of zebrafish larvae is decreased in the presence of a weak prepulse. In addition, the administration of the dopamine agonist apomorphine in the medium suppresses PPI. However, the pretreatment of the larvae with the anti-psychotic haloperidol inhibits the apomorphine effect. To identify genes that regulate PPI, a mutagenesis screen was performed using ethylnitrosourea (ENU) as a mutagen. They screened 686 genomes and identify five mutant lines with reduced PPI. The establishment of this assay will allow for additional screens to identify genes involved in sensory motor gating. In addition, it will allow for the characterization of transgenic and mutant zebrafish lines of schizophrenia susceptibility genes.

Zebrafish as a model to determine susceptibility gene function

Zebrafish have been used to study the function of schizophrenia susceptibility genes (Drerup et al., 2009; Wood et al., 2009). Wood et al. (2009) examined the function of *disc1* and *neuregulin 1* in oligodendrocyte and neuron specification in the zebrafish brain. They determined that knock down of Disc1 and Neuregulin 1 in zebrafish embryos using MOs resulted in defects in oligodendrocyte development and loss of *olig2*-positive cerebellar neurons. Their results suggest a role for *disc1* and *neuregulin 1* in the development of oligodendrocytes and neurons from *olig2*-precursor cells.

We determined that *disc1* is expressed in cranial neural crest (CNC) cells by fluorescent in situ hybridization in the Tg(*−4.9sox10:egfp*)^{ba2} zebrafish transgenic line (Drerup et al., 2009). CNC cells are multipotent progenitors that delaminate from the ectoderm covering the dorsal neural tube. These cells migrate in streams to produce multiple cell types including craniofacial cartilage, pigment and neurons and glia of the peripheral nervous system (Knight and Schilling, 2006). Of importance to our studies into Disc1 function, neural crest and neural cells of the developing brain share many features including their ectodermal origin, ability to migrate long distances, the ability to give rise to multiple cell types, and their responsiveness to the same intracellular and extracellular signaling molecules (Lefcort et al., 2007). Therefore, understanding the function of Disc1 in this well-characterized and easily observable cell population may provide valuable insights into the function of this protein in the developing brain.

We determined using a Disc1 spice-blocking MO (MOE3) (Fig. 1) that Disc1 knockdown results in medial expansion of the CNC markers, *foxd3* and *sox10*, in premigratory CNC cells (Fig. 2) (Drerup et al., 2009). This data indicates that Disc1 plays a role in the transcriptional repression of these

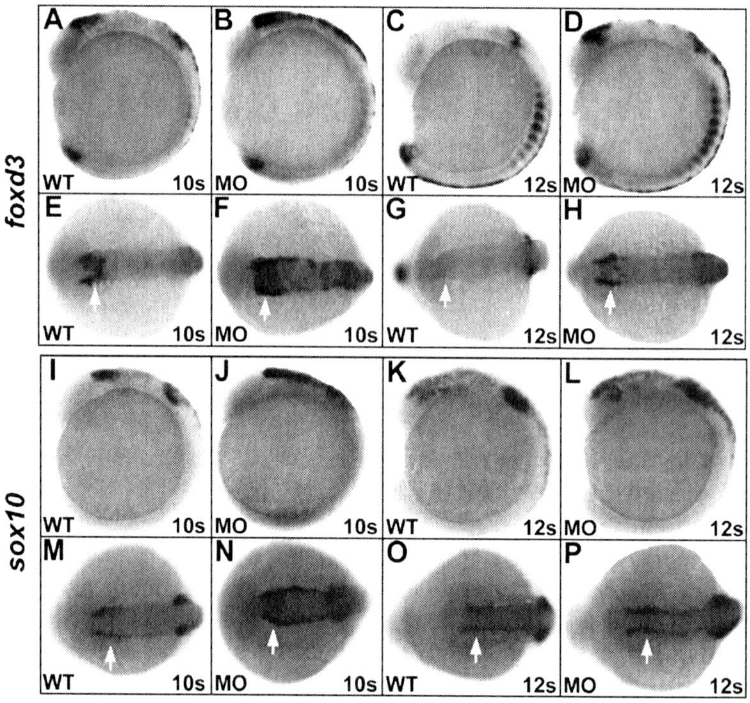

Fig. 2. Medial expansion of *foxd3* and *sox10* expression in Disc1 morphants. In situ hybridization analysis at 10 s (somites) and 12 s demonstrates a medial expansion and increased level of *foxd3* (A–H) and *sox10* (I–P) expression compared to wild-type control embryos (white arrows). Lateral views (A–D, I–L) and dorsal views (E–H, M–P) are presented with anterior to the left. The figure is modified from Drerup et al. (2009).

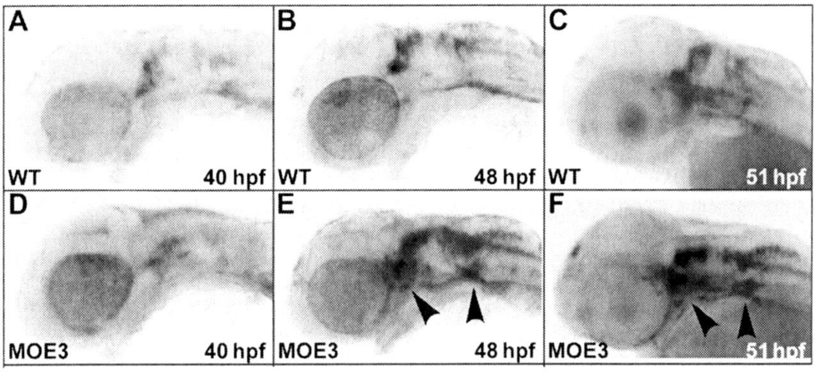

G

Embryo Type	Age	n	% Glial Expansion
WT	40 hpf	17	0%
MOE3	40 hpf	50	0%
WT	48 hpf	18	0%
MOE3	48 hpf	125	88%
WT	51 hpf	21	0%
MOE3	51 hpf	175	87%

Fig. 3. Zebrafish embryos with Disc1 knockdown have expanded peripheral cranial glia populations. (A–F) In situ hybridization analysis demonstrated that *foxd3* expression was expanded at 48 and 51 hpf in Disc1 morphants. At an earlier timepoint (40 hpf), *foxd3* expression is not expanded in either Disc1 morphants or controls. At these timepoints, *foxd3* marks developing glia populations. The arrowheads indicate enhanced expression in the trigeminal ganglion and posterior lateral line ganglion. (G) Quantification of the glial expansion in Disc1 morphants compared to controls. The figure is modified from Drerup et al. (2009).

transcription factors that have critical functions in CNC cells including the maintenance of progenitor pools, regulating migration onset and the differentiation of derivatives. Using time-lapse imaging in Disc1 morphant embryos, we then monitored the behavior of migrating CNC cells and determined that loss of Disc1 results in hindered migration. Furthermore, we determined that the continued expression of *sox10* in Disc1 morphants resulted in a reduction in CNC cell populations in the pharyngeal arches resulting in craniofacial defects and an expansion of the peripheral cranial glia populations (Fig. 3) indicating a potential change in cell fate. Both Foxd3 and Sox10 play roles in brain development. Foxd3 is a stem cell marker that is present in neurogenic brain regions (Lein et al., 2007) and Sox10 plays a critical role in oligodendrocyte differentiation (Wegner and Stolt, 2005). Moreover, oligodendrocyte dysfunction has been implicated in the pathogenesis of schizophrenia (Hoistad et al., 2009). Using the zebrafish model system, we were able to identify a unique Disc1 function that may play a critical role in cell migration, fate determination and differentiation in the developing brain.

Acknowledgments

I would like to thank Kate Meyer and Heather Wiora for their critical reading of this review. In addition, I would like to thank Catherine Drerup, Heather Wiora and Jacek Topczewski for their collaboration on researching Disc1 function in zebrafish.

References

Abraham, E., Palevitch, O., Gothilf, Y., & Zohar, Y. (2009). The zebrafish as a model system for forebrain GnRH neuronal development. *General and Comparative Endocrinology,* doi: 10.1016/j.ygcen.2009.01.012. 30 January 2009. [Epub ahead of print].

Amatruda, J. F., & Patton, E. E. (2008). Genetic models of cancer in zebrafish. *International Review of Cell and Molecular Biology*, *271*, 1–34.

Amatruda, J. F., Shepard, J. L., Stern, H. M., & Zon, L. I. (2002). Zebrafish as a cancer model system. *Cancer Cell*, *1*, 229–231.

Amsterdam, A., & Hopkins, N. (2006). Mutagenesis strategies in zebrafish for identifying genes involved in development and disease. *Trends in Genetics*, *22*, 473–478.

Arguello, P. A., & Gogos, J. A. (2006). Modeling madness in mice: One piece at a time. *Neuron*, *52*, 179–196.

Beattie, C. E., Carrel, T. L., & Mcwhorter, M. L. (2007). Fishing for a mechanism: Using zebrafish to understand spinal muscular atrophy. *Journal of Child Neurology*, *22*, 995–1003.

Bilder, R. M., Volavka, J., Czobor, P., Malhotra, A. K., Kennedy, J. L., Ni, X., et al. (2002). Neurocognitive correlates of the COMT Val(158)Met polymorphism in chronic schizophrenia. *Biological Psychiatry*, *52*, 701–707.

Bill, B. R., Petzold, A. M., Clark, K. J., Schimmenti, L. A., & Ekker, S. C. (2009). A primer for morpholino use in zebrafish. *Zebrafish*, *6*, 69–77.

Brandon, N. J., Handford, E. J., Schurov, I., Rain, J. C., Pelling, M., Duran-Jimeniz, B., et al. (2004). Disrupted in Schizophrenia 1 and Nudel form a neurodevelopmentally regulated protein complex: Implications for schizophrenia and other major neurological disorders. *Molecular and Cellular Neuroscience*, *25*, 42–55.

Burgess, H. A., & Granato, M. (2007a). Modulation of locomotor activity in larval zebrafish during light adaptation. *Journal of Experimental Biology*, *210*, 2526–2539.

Burgess, H. A., & Granato, M. (2007b). Sensorimotor gating in larval zebrafish. *Journal of Neuroscience*, *27*, 4984–4994.

Burgess, H. A., & Granato, M. (2008). The neurogenetic frontier — lessons from misbehaving zebrafish. *Briefings in Functional Genomic & Proteomic*, *7*, 474–482.

Chico, T. J., Ingham, P. W., & Crossman, D. C. (2008). Modeling cardiovascular disease in the zebrafish. *Trends in Cardiovascular Medicine*, *18*, 150–155.

Darland, T., & Dowling, J. E. (2001). Behavioral screening for cocaine sensitivity in mutagenized zebrafish. *Proceedings of the National Academy of Science of the United States of America*, *98*, 11691–11696.

Darrow, K. O., & Harris, W. A. (2004). Characterization and development of courtship in zebrafish, *Danio rerio*. *Zebrafish*, *1*, 40–45.

Davidson, A. E., Balciunas, D., Mohn, D., Shaffer, J., Hermanson, S., Sivasubbu, S., et al. (2003). Efficient gene delivery and gene expression in zebrafish using the Sleeping Beauty transposon. *Developmental Biology*, *263*, 191–202.

Deiters, A., & Yoder, J. A. (2006). Conditional transgene and gene targeting methodologies in zebrafish. *Zebrafish*, *3*, 415–429.

Doyon, Y., Mccammon, J. M., Miller, J. C., Faraji, F., Ngo, C., Katibah, G. E., et al. (2008). Heritable targeted gene disruption in zebrafish using designed zinc-finger nucleases. *Nature Biotechnology*, *26*, 702–708.

Draper, B. W., Morcos, P. A., & Kimmel, C. B. (2001). Inhibition of zebrafish fgf8 pre-mRNA splicing with morpholino oligos: A quantifiable method for gene knockdown. *Genesis*, *30*, 154–156.

Drerup, C. M., Ahlgren, S. C., & Morris, J. A. (2007). Expression profiles of ndel1a and ndel1b, two orthologs of the NudE-like gene, in the zebrafish. *Gene Expression Patterns*, *7*, 672–679.

Drerup, C. M., Wiora, H. M., Topczewski, J., & Morris, J. A. (2009). Disc1 regulates foxd3 and sox10 expression, affecting neural crest migration and differentiation. *Development*, *136*, 2623–2632.

Eaton, R. C., Farley, R. D., Kimmel, C. B., & Schabtach, E. (1977). Functional development in the Mauthner cell system of embryos and larvae of the zebra fish. *Journal of Neurobiology*, *8*, 151–172.

Efimov, V. P., & Morris, N. R. (2000). The LIS1-related NUDF protein of *Aspergillus nidulans* interacts with the coiled-coil domain of the NUDE/RO11 protein. *Journal of Cell Biology*, *150*, 681–688.

Egan, M. F., Goldberg, T. E., Kolachana, B. S., Callicott, J. H., Mazzanti, C. M., Straub, R. E., et al. (2001). Effect of COMT Val108/158 Met genotype on frontal lobe function and risk for schizophrenia. *Proceedings of the National Academy of Science of the United States of America*, *98*, 6917–6922.

Engeszer, R. E., Ryan, M. J., & Parichy, D. M. (2004). Learned social preference in zebrafish. *Current Biology*, *14*, 881–884.

Fleisch, V. C., & Neuhauss, S. C. (2006). Visual behavior in zebrafish.. *Zebrafish*, *3*, 191–201.

Gerlai, R. (2003). Zebra fish: An uncharted behavior genetic model. *Behavior Genetics*, *33*, 461–468.

Gerlai, R., Fernandes, Y., & Pereira, T. (2009). Zebrafish (*Danio rerio*) responds to the animated image of a predator: Towards the development of an automated aversive task. *Behavioural Brain Research*, *201*, 318–324.

Gottesman, Ii, & Erlenmeyer-Kimling, L. (2001). Family and twin strategies as a head start in defining prodromes and endophenotypes for hypothetical early-interventions in schizophrenia. *Schizophrenia Research*, *51*, 93–102.

Grandel, H., Kaslin, J., Ganz, J., Wenzel, I., & Brand, M. (2006). Neural stem cells and neurogenesis in the adult zebrafish brain: Origin, proliferation dynamics, migration and cell fate. *Developmental Biology*, *295*, 263–277.

Grunwald, D. J., & Eisen, J. S. (2002). Headwaters of the zebrafish — emergence of a new model vertebrate. *Nature Reviews Genetics*, *3*, 717–724.

Harvey, R. P. A. D., & Hoistad, M. (2002). Schizophrenia: Course over a lifetime. In K. L. Kavis, D. Charney, J. T. Coyle, & C. Nemeroff (Eds.), *Neuropsychopharmacology: The fifth generation of progress*.

Hatta, K., Tsujii, H., & Omura, T. (2006). Cell tracking using a photoconvertible fluorescent protein. *Nature Protocols*, *1*, 960–967.

Hoistad, M., Segal, D., Takahashi, N., Sakurai, T., Buxbaum, J. D., & Hof, P. R. (2009). Linking white and grey matter in schizophrenia: Oligodendrocyte and neuron pathology in the prefrontal cortex. *Frontiers in Neuroanatomy*, *3*, 9.

Kamiya, A., Tomoda, T., Chang, J., Takaki, M., Zhan, C., Morita, M., et al. (2006). DISC1-NDEL1/NUDEL protein interaction, an essential component for neurite outgrowth, is modulated by genetic variations of DISC1. *Human Molecular Genetics*, *15*, 3313–3323.

Kari, G., Rodeck, U., & Dicker, A. P. (2007). Zebrafish: An emerging model system for human disease and drug discovery. *Clinical Pharmacology and Therapeutics*, *82*, 70–80.

Kawakami, K., Takeda, H., Kawakami, N., Kobayashi, M., Matsuda, N., & Mishina, M. (2004). A transposon-mediated gene trap approach identifies developmentally regulated genes in zebrafish. *Developmental Cell*, *7*, 133–144.

Kimmel, C. B., Patterson, J., & Kimmel, R. O. (1974). The development and behavioral characteristics of the startle response in the zebra fish. *Developmental Psychobiology*, *7*, 47–60.

Knight, R. D., Nair, S., Nelson, S. S., Afshar, A., Javidan, Y., Geisler, R., et al. (2003). lockjaw encodes a zebrafish tfap2a required for early neural crest development. *Development*, *130*, 5755–5768.

Knight, R. D., & Schilling, T. F. (2006). Cranial neural crest and development of the head skeleton. *Advances in Experimental Medicine and Biology*, *589*, 120–133.

Lam, C. S., Marz, M., & Strahle, U. (2009). gfap and nestin reporter lines reveal characteristics of neural progenitors in the adult zebrafish brain. *Developmental Dynamics*, *238*, 475–486.

Lefcort, F., O'connor, T., & Kulesa, P. M. (2007). Comparative analysis of neural crest cell and axonal growth cone dynamics and behavior. In I. De Curtis (Ed.), *Intracellular mechanisms for neuritogenesis*. New York: Springer.

Lein, E. S., Hawrylycz, M. J., Ao, N., Ayres, M., Bensinger, A., Bernard, A., et al. (2007). Genome-wide atlas of gene expression in the adult mouse brain. *Nature*, *445*, 168–176.

Lewis, D. A., & Levitt, P. (2002). Schizophrenia as a disorder of neurodevelopment. *Annual Review of Neuroscience*, *25*, 409–432.

Lewis, D. A., & Lieberman, J. A. (2000). Catching up on schizophrenia: Natural history and neurobiology. *Neuron*, *28*, 325–334.

Lipska, B. K., & Weinberger, D. R. (2000). To model a psychiatric disorder in animals: Schizophrenia as a reality test. *Neuropsychopharmacology*, *23*, 223–239.

Marenco, S., & Weinberger, D. R. (2000). The neurodevelopmental hypothesis of schizophrenia: Following a trail of evidence from cradle to grave. *Development of Psychopathology*, *12*, 501–527.

Meng, X., Noyes, M. B., Zhu, L. J., Lawson, N. D., & Wolfe, S. A. (2008). Targeted gene inactivation in zebrafish using engineered zinc-finger nucleases. *Nature Biotechnology*, *26*, 695–701.

Millar, J. K., Christie, S., & Porteous, D. J. (2003). Yeast two-hybrid screens implicate DISC1 in brain development and function. *Biochemical and Biophysical Research Communications*, *311*, 1019–1025.

Millar, J. K., Wilson-Annan, J. C., Anderson, S., Christie, S., Taylor, M. S., Semple, C. A., et al. (2000). Disruption of two novel genes by a translocation co-segregating with schizophrenia. *Human Molecular Genetics*, *9*, 1415–1423.

Miller, N., & Gerlai, R. (2007). Quantification of shoaling behaviour in zebrafish (*Danio rerio*). *Behavioural Brain Research*, *184*, 157–166.

Mione, M., Baldessari, D., Deflorian, G., Nappo, G., & Santoriello, C. (2008). How neuronal migration contributes to the morphogenesis of the CNS: Insights from the zebrafish. *Developmental Neuroscience*, *30*, 65–81.

Morris, J. A., Kandpal, G., Ma, L., & Austin, C. P. (2003). DISC1 (Disrupted-In-Schizophrenia 1) is a centrosome-associated protein that interacts with MAP1A, MIPT3, ATF4/5 and NUDEL: Regulation and loss of interaction with mutation. *Human Molecular Genetics*, *12*, 1591–1608.

Nasevicius, A., & Ekker, S. C. (2000). Effective targeted gene "knockdown" in zebrafish. *Nature Genetics*, *26*, 216–220.

Newman, M., Musgrave, I. F., & Lardelli, M. (2007). Alzheimer disease: Amyloidogenesis, the presenilins and animal models. *Biochimica et Biophysics Acta*, *1772*, 285–297.

Ozeki, Y., Tomoda, T., Kleiderlein, J., Kamiya, A., Bord, L., Fujii, K., et al. (2003). Disrupted-in-Schizophrenia-1 (DISC-1): Mutant truncation prevents binding to NudE-like (NUDEL) and inhibits neurite outgrowth. *Proceedings of the National Academy of Science of the United States of America*, *100*, 289–294.

Powell, S. B., Zhou, X. & Geyer, M. A. (2009) Prepulse inhibition and genetic mouse models of schizophrenia. Behavioural Brain Research, Epub ahead of print.

Rink, E., & Wullimann, M. F. (2004). Connections of the ventral telencephalon (subpallium) in the zebrafish (*Danio rerio*). *Brain Research*, *1011*, 206–220.

Robu, M. E., Larson, J. D., Nasevicius, A., Beiraghi, S., Brenner, C., Farber, S. A., et al. (2007). p53 activation by knockdown technologies. *PLoS Genetics*, *3*, e78.

Rodriguez, F., Lopez, J. C., Vargas, J. P., Broglio, C., Gomez, Y., & Salas, C. (2002a). Spatial memory and hippocampal pallium through vertebrate evolution: Insights from reptiles and teleost fish. *Brain Research Bulletin*, *57*, 499–503.

Rodriguez, F., Lopez, J. C., Vargas, J. P., Gomez, Y., Broglio, C., & Salas, C. (2002b). Conservation of spatial memory function in the pallial forebrain of reptiles and ray-finned fishes. *Journal of Neuroscience*, *22*, 2894–2903.

Russek-Blum, N., Nabel-Rosen, H., & Levkowitz, G. (2009). High resolution fate map of the zebrafish diencephalon. *Developmental Dynamics*, *238*, 1827–1835.

Sawa, A., & Snyder, S. H. (2002). Schizophrenia: Diverse approaches to a complex disease. *Science*, *296*, 692–695.

Stefansson, H., Sarginson, J., Kong, A., Yates, P., Steinthorsdottir, V., Gudfinnsson, E., et al. (2003). Association of neuregulin 1 with schizophrenia confirmed in a Scottish population. *American Journal of Human Genetics*, *72*, 83–87.

Stefansson, H., Sigurdsson, E., Steinthorsdottir, V., Bjornsdottir, S., Sigmundsson, T., Ghosh, S., et al. (2002). Neuregulin 1 and susceptibility to schizophrenia. *American Journal of Human Genetics*, *71*, 877–892.

Straub, R. E., Jiang, Y., Maclean, C. J., Ma, Y., Webb, B. T., Myakishev, M. V., et al. (2002). Genetic variation in the 6p22.3 gene DTNBP1, the human ortholog of the mouse dysbindin gene, is associated with schizophrenia. *American Journal of Human Genetics*, *71*, 337–348.

Sullivan, P. F., Kendler, K. S., & Neale, M. C. (2003). Schizophrenia as a complex trait: Evidence from a meta-analysis of twin studies. *Archivesof General Psychiatry*, *60*, 1187–1192.

Taylor, J. S., Braasch, I., Frickey, T., Meyer, A., & Van De Peer, Y. (2003). Genome duplication, a trait shared by 22000 species of ray-finned fish. *Genome Research*, *13*, 382–390.

Tropepe, V., & Sive, H. L. (2003). Can zebrafish be used as a model to study the neurodevelopmental causes of autism? *Genes Brain and Behavior*, *2*, 268–281.

Vitebsky, A., Reyes, R., Sanderson, M. J., Michel, W. C., & Whitlock, K. E. (2005). Isolation and characterization of the laure olfactory behavioral mutant in the zebrafish, *Danio rerio*. *Developmental Dynamics*, *234*, 229–242.

Wegner, M., & Stolt, C. C. (2005). From stem cells to neurons and glia: A Soxist's view of neural development. *Trends in Neurosciences*, *28*, 583–588.

Wood, J. D., Bonath, F., Kumar, S., Ross, C. A., & Cunliffe, V. T. (2009). Disrupted-in-schizophrenia 1 and neuregulin 1 are required for the specification of oligodendrocytes and neurones in the zebrafish brain. *Human Molecular Genetics*, *18*, 391–404.

Woods, I. G., Wilson, C., Friedlander, B., Chang, P., Reyes, D. K., Nix, R., et al. (2005). The zebrafish gene map defines ancestral vertebrate chromosomes. *Genome Research*, *15*, 1307–1314.

Wullimann, M. F., & Mueller, T. (2004). Teleostean and mammalian forebrains contrasted: Evidence from genes to behavior. *Journal of Comparative Neurology*, *475*, 143–162.

Wullimann, M. F., & Rink, E. (2002). The teleostean forebrain: A comparative and developmental view based on early proliferation, Pax6 activity and catecholaminergic organization. *Brain Research Bulletin*, *57*, 363–370.

Wynshaw-Boris, A. (2007). Lissencephaly and LIS1: Insights into the molecular mechanisms of neuronal migration and development. *Clinical Genetics*, *72*, 296–304.

Zupanc, G. K. (2008). Adult neurogenesis and neuronal regeneration in the brain of teleost fish. *Journal of Physiology Paris*, *102*, 357–373.

A. Sawa (Ed.)
Progress in Brain Research, Vol. 179
ISSN 0079-6123

CHAPTER 12

Modeling schizophrenia in flies

Katsuo Furukubo-Tokunaga*

Graduate School of Life and Environmental Sciences, University of Tsukuba, Tennodai, Tsukuba, Japan

Abstract: Schizophrenia is a debilitating mental illness that affects 1% of the population worldwide. Although its molecular etiology remains unclear, recent advances in human psychiatric genetics have identified a large number of candidate genetic risk factors involved in schizophrenia. Modeling the disease in genetically tractable animals is thus a challenging but increasingly important task. In this review, I discuss the potential problems and perspectives associated with modeling schizophrenia in fruit flies, and briefly review the recent studies analyzing the molecular and cellular functions of Disrupted-In-Schizophrenia-1 (DISC1) in transgenic flies.

Keywords: schizophrenia; genetic model; *Drosophila*; sleep; cAMP; ATF; mushroom body

Introduction

Although *Drosophila melanogaster* has been used successfully as a tool to study human genetics (Bonini and Fortini, 2003; Bilen and Bonini, 2005; Lessing and Bonini, 2009), it should be acknowledged that there are limitations in fly models as in other animal models, particularly with regard to the processes that are considered only manifested in humans. For example, genes involved in creating the four-chambered heart or in elaborating the coordinated system of ducts in mammary glands would not be ideal targets to examine in fruit flies (Bier, 2005). However, genes controlling each of the specific steps involved in these processes might have conserved functions, and their function within those pathways could be studied in flies. Likewise, in the case of schizophrenia, flies would not be an ideal system to model delusions, hallucinations, thought distortions and other complex mental processes that are specifically manifested in psychiatric patients. Nevertheless, genes involved in basic biological processes that relate to endophenotypes of the disease (Gottesman and Gould, 2003), including learning and memory (Margulies et al., 2005; McGuire et al., 2005; Berry et al., 2008), attention (van Swinderen and Greenspan, 2003; van Swinderen, 2005; van Swinderen, 2006), sleep and arousal (Hendricks et al., 2000; Shaw et al., 2000; Andretic and Shaw, 2005; Ho and Sehgal, 2005; Zimmerman et al., 2008; Harbison et al., 2009), as well as intracellular molecular cascades such as cAMP signaling pathways (Millar et al., 2007; Chubb et al., 2008; Hains and Arnsten, 2008), which are conserved in both vertebrates and invertebrates and have pivotal roles in regulating

*Corresponding author.
Tel.: +8 129 853 6644; Fax: +8 129 853 6644;
E-mail: furukubo-tokunaga.gm@u.tsukuba.ac.jp

DOI: 10.1016/S0079-6123(09)17912-8

the discrete processes of these endogenous phenotypes could be studied in flies. Information on the interactions between conserved components of the molecular pathways in flies could then help to identify novel human genes and gene pathways as well as the cellular processes that are controlled by the homologous genes and underlie cognitive and mental deficits in patients. In this brief review, I discuss the versatility and limitations of the *Drosophila* system in modeling schizophrenia with respect to recently developed fly genetic tools and their application to the molecular study of neuronal functions of a susceptibility gene.

Drosophila as a model for cellular and molecular studies of susceptibility genes

Although a number of animal models have been used to examine different pathophysiological aspects of schizophrenia (Arguello and Gogos, 2006), flies are most suitable for examining cellular and molecular processes, which occur upstream of the observed clinical psychopathology. The complete genome sequence of *Drosophila melanogaster* has been determined as early as in 2000 (Adams et al., 2000), and direct comparison of the genomic sequences demonstrates that a large number of genes are conserved between flies and humans (Venter et al., 2001), making *Drosophila* as an ideal model system for the study of molecular and cellular functions. Such an approach would help to identify the cellular and synaptic substrates underlying the developmental or functional disruptions observed in schizophrenic patients, and to identify molecular and genetic components interacting with the gene products, resulting in the elucidation of gene pathways and cellular processes involved in psychological disorders. The outcomes can then be verified in follow-up studies in mammalian models and human genetic studies that may eventually lead to novel therapeutic strategies. The powerful genetic techniques available in *Drosophila* can be exploited to identify candidate genes that are relevant to psychological disorders but might have been previously overlooked due to the complexity and limitations of the mammalian models and human genetic studies.

Genetics in flies

The forward genetic approach, in which mutants are selected and analyzed based on their loss-of-function phenotypes, has been widely used in flies and has proven to be highly rewarding. Using this approach, a number of studies have revealed both normal and aberrant functions of human disease genes in living organisms (Bonini and Fortini, 2003; Bier, 2005; Bilen and Bonini, 2005; Lessing and Bonini, 2009). Due to the development of the systematic gene disruption project (Bier, 2005), loss-of-function mutations are available for the majority of the *Drosophila melanogaster* genes, and the existence of the highly integrated database Fly Base (Tweedie et al., 2009) aids identification of cellular and molecular phenotypes of the mutations, and even the availability of the corresponding mutant stocks in stock centers. Moreover, recent generation of a genome-wide library of *Drosophila melanogaster* RNAi transgenes (Dietzl et al., 2007) enables conditional inactivation of gene function in specific neurons of the intact organism at specific developmental stages.

Although the forward genetic approach is highly useful for studying the molecular functions of the homologous genes in flies, it cannot be used for those genes that are not conserved in flies. Additionally, because of the diversity of gene functions, this approach does not always lead to direct molecular parallels or insights that are immediately applicable to human disease (Bonini and Fortini, 2003; Bilen and Bonini, 2005; Lessing and Bonini, 2009). An alternative approach, whereby the human disease genes are directly expressed in fruit flies, has proven highly useful in analyzing the resulting cellular phenotypes and for elucidating genetic modifier pathways involved in neurodegenerative diseases. The vast arrays of transgenic techniques in *Drosophila melanogaster* allow us to drive the expression of an exogenous genetic construct in specific sets of neurons and developmental stages in living animals. This direct expression approach can be used for overexpression or misexpression of either wild-type or mutated human genes in flies of various genetic backgrounds including those carrying loss-of-function mutations with related susceptibility genes. A number of techniques are available for the spatially or temporally controlled expression of the human transgenes in the fly brain (McGuire et al., 2004), which involves the use of either defined promoter sequences or transcriptional enhancer trapping methods in conjunction with the GAL4-UAS system (Brand and Perrimon, 1993).

This directed expression approach is distinct from the classical forward genetic approach, and allows straightforward analysis of the cellular and molecular functions of the human gene products and the identification of novel interacting genes that functions with the transgenes in conserved pathways involved in the cellular and molecular alterations. Furthermore, it is feasible to screen a large number of genes for novel factors that either enhance or suppress the phenotypes caused by the expression of the human transgenes (Bonini and Fortini, 2003; Bilen and Bonini, 2005). In many cases, such interacting factors found in flies are found to interact with disease proteins in mammalian systems and indeed participate in the pathological pathways. Thus, even if disease genes by themselves are not conserved in flies, transgenic flies expressing the human-specific genes often exhibit unique and intriguing phenotypes via protein–protein interactions of the exogenously expressed disease proteins and their endogenous interactors that are conserved between flies and humans. This has been demonstrated in several studies on neurodegenerative diseases including Parkinson's disease (Feany and Bender, 2000; Auluck et al., 2002). For example, although there is no direct orthologue of the α-synuclein gene in flies, the fly model expressing the human α-synuclein gene exhibits adult-onset loss of specific sets of dopaminergic neurons, similar to human Parkinson's disease, and has helped to identify novel factors including Hsp70 (Auluck et al., 2002; Wu et al., 2004), which was later confirmed as a modified factor or genetic risk factor for Parkinson's disease (Wu et al., 2004).

DISC1 analyses in flies

Disrupted-In-Schizophrenia-1 (DISC1) is a promising genetic factor for a wide range of mental illnesses, including schizophrenia, bipolar disorder, major depression, and autism spectrum conditions (Harrison and Weinberger, 2005; Ishizuka et al., 2006; Chubb et al., 2008; Kilpinen et al., 2008; Jaaro-Peled et al., 2009). DISC1 was originally identified as the gene located on chromosome 1 and interrupted by a balanced translocation t(1q42.1;11q14.3) that is linked to psychopathology including schizophrenia, depression, and bipolar disorder in a large Scottish family (St Clair et al., 1990; Millar et al., 2000). Cellular and molecular studies have revealed that the DISC1 protein has multiple functions and is located in several distinct domains in cells, including the centrosome, the postsynaptic density, the mitochondria, and the nucleus (Millar et al., 2005; Kirkpatrick et al., 2006). DISC1 has specific protein interactors, which correspond to each subcellular domain, including NudE-like (NUDEL/NDEL1) at the centrosome and activating transcription factor 4 (ATF4)/CREB2 in the nucleus (Morris et al., 2003; Kamiya et al., 2005). In addition, a nuclear form of DISC1 is enriched in patients with sporadic schizophrenia and major depression, suggesting that DISC1 may play a role in transcriptional control of related genes involved in the pathophysiology of schizophrenia and bipolar disorders (Sawamura et al., 2005).

Recently, a *Drosophila* model for examining cellular and molecular functions of DISC1 in living organisms was developed (Sawamura et al., 2008). In this study, the full-length form and c-terminally truncated form (amino acid residue 1-597 corresponding the disrupted form by the 1q42.1;11q14.3 translocation) were expressed in *Drosophila* mushroom bodies, centers of cognitive functions, such as associative learning, attention, and sleep/arousal control in the fly brain (Heisenberg, 2003; van Swinderen et al., 2004; Joiner et al., 2006; Pitman et al., 2006), using an inducible gene expression system, GeneSwitch (Mao et al., 2004), which allows precise targeting of DISC1 expression to specific sets of neurons in the fly brain of specific developmental stages (Fig. 1). The full-length DISC1 exhibited punctate localization in the nuclei of the fly neurons, while truncated DISC1 exhibited a diffuse expression pattern in both the

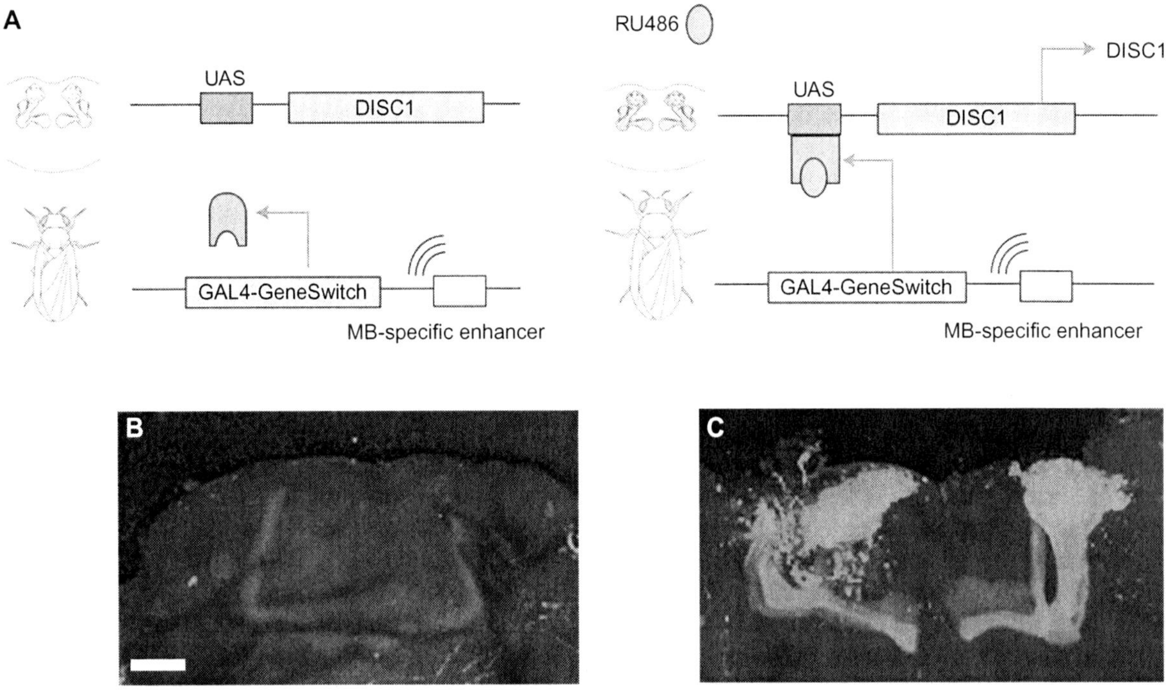

Fig. 1. Cell-type and stage-specific expression of DISC1 in flies. (A) The GAL4-GeneSwitch chimeric protein binds to a UAS-DNA sequence in the presence of RU486, which in turn induces gene transcription downstream of the UAS sequences. Inducible expression of DISC1 selectively in mushroom body (MB) is achieved by using an MB-GeneSwitch construct, in which a MB-specific enhancer is linked to the coding region of GAL4-GeneSwitch. (B, C) Confocal images of adult fly brains expressing the mCD8::GFP protein under control of the MB-GeneSwitch. In the presence of RU486 (B, 0 μM; C, 250 μM), selective expression of the protein is observed in mushroom bodies.

nuclei and cytoplasm (Fig. 2). Neither the full-length nor the truncated form of DISC1 caused gross anatomical alterations of mushroom body structures, which retained characteristic expression of molecular markers including Fas II, a cell adhesion molecule of IgG super family. By contrast, transgenic flies expressing the full-length DISC1 display disturbance in sleep homeostasis (Fig. 3), which is controlled by CREB signaling and CRE-mediated gene transcription (Hendricks et al., 2001; Zimmerman et al., 2008). Flies are known to exhibit a typical circadian activity pattern with more frequent and longer rest at night than during daytime; this can be measured by using a computerized activity monitor system (Drosophila Activity Monitor System) that counts the number of infrared beam crossings by an individual fly placed in a small glass tube (Fig. 3). Quantitative characteristics measured with this system reflect the arousal and sleep homeostasis of flies (Hendricks et al., 2000; Shaw et al., 2000; Andretic and Shaw, 2005; Ho and Sehgal, 2005; Zimmerman et al., 2008; Harbison et al., 2009;). Targeted expression of the full-length DISC1 to mushroom body neurons failed to affect sleep in female flies but stimulated the amount of sleep in male flies (Fig. 3). By contrast, the c-terminally truncated DISC1 did not affect sleep in male or female flies although a slight increase of the daytime sleep was noticeable in males (Fig. 3). Of note, DISC1 does not alter the circadian cycle or the locomotor activity of the transgenic flies. Rather, the increased sleep in male flies expressing the full-length DISC1 is caused by prolonged duration of each sleep period, whereas the number of total sleep events in 24 h is unaltered in the transgenic flies (Sawamura et al., 2008).

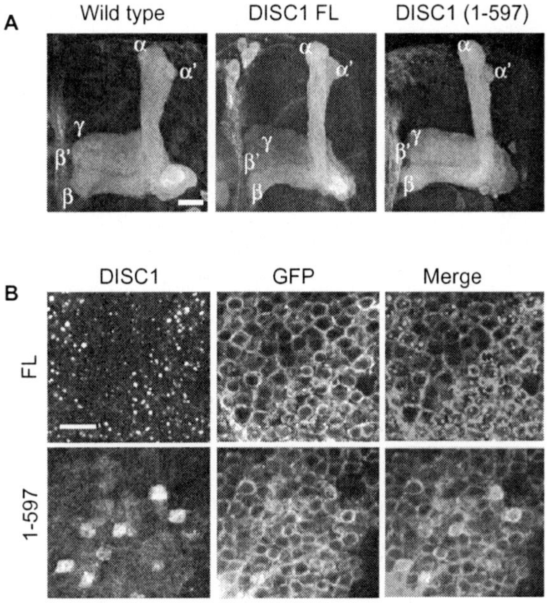

Fig. 2. DISC1 expression in *Drosophila* mushroom body neurons. (A) Gross morphology of mushroom bodies is unaltered by the expression of DISC1. Note that the characteristic lobe systems are normally formed in DISC1-FL and DISC1(1-597) flies. FASII, which is expressed in α and β lobes, is also unchanged in the DISC1 flies. DISC1 expression was driven by OK107, a mushroom body Gal4 driver. Lobes were visualized with UAS-mCD8::GFP and anti-FASII staining. Scale bar, 10 μm. (B) Localization of DISC1 in *Drosophila* neurons. Note that the DISC1-FL exhibits punctate nuclear localization, whereas DISC1(1-597) localizes diffusely both in the nuclei and cytoplasm. Scale bar, 10 μm.

Sleep homeostasis in flies is associated with CREB signaling and CRE-mediated gene transcription (Hendricks et al., 2001; Zimmerman et al., 2008), consistent with the observation that DISC1 directly interacts with ATF4/CREB2 transcription factors and a corepressor N-CoR in vitro (Sawamura et al., 2008), leading to the modulation of CRE-mediated gene transcription in cultured cells. As is the case in *Drosophila* neurons, DISC1 is localized in the nucleus of mammalian cortical neurons and HeLa cells, where punctate DISC1 signals co-localize with PML bodies, nuclear compartments for gene transcription. Intriguingly, molecular studies have also demonstrated a direct interaction between the human DISC1 and the *Drosophila* Cryptocephal protein, the fly orthologue of the human ATF4/CREB2, implying a conserved molecular network of DISC1 interacting partners in the *Drosophila* neurons (Sawamura et al., 2008).

Perspectives

The recent progress of psychiatric genetics and genome-wide association studies have disclosed a large number of genes affecting the susceptibility for schizophrenia and related psychiatric disorders (O'Donovan et al., 2008; Walsh et al., 2008). In parallel, recent behavioral and neurogenetical studies in *Drosophila* have shown surprising richness of fly's cognitive abilities (Greenspan and van Swinderen, 2004). *Drosophila* is now known to perform a variety of sophisticated cognitive tasks, such as context generalization and position invariance (Liu et al., 1999; Tang et al., 2004; Neuser et al., 2008), not to mention other complex behavioral trains including associative learning (Margulies et al., 2005; McGuire et al., 2005; Berry et al., 2008), attention (van Swinderen and Greenspan, 2003; van Swinderen, 2005; van Swinderen, 2006), and arousal (Hendricks et al.,

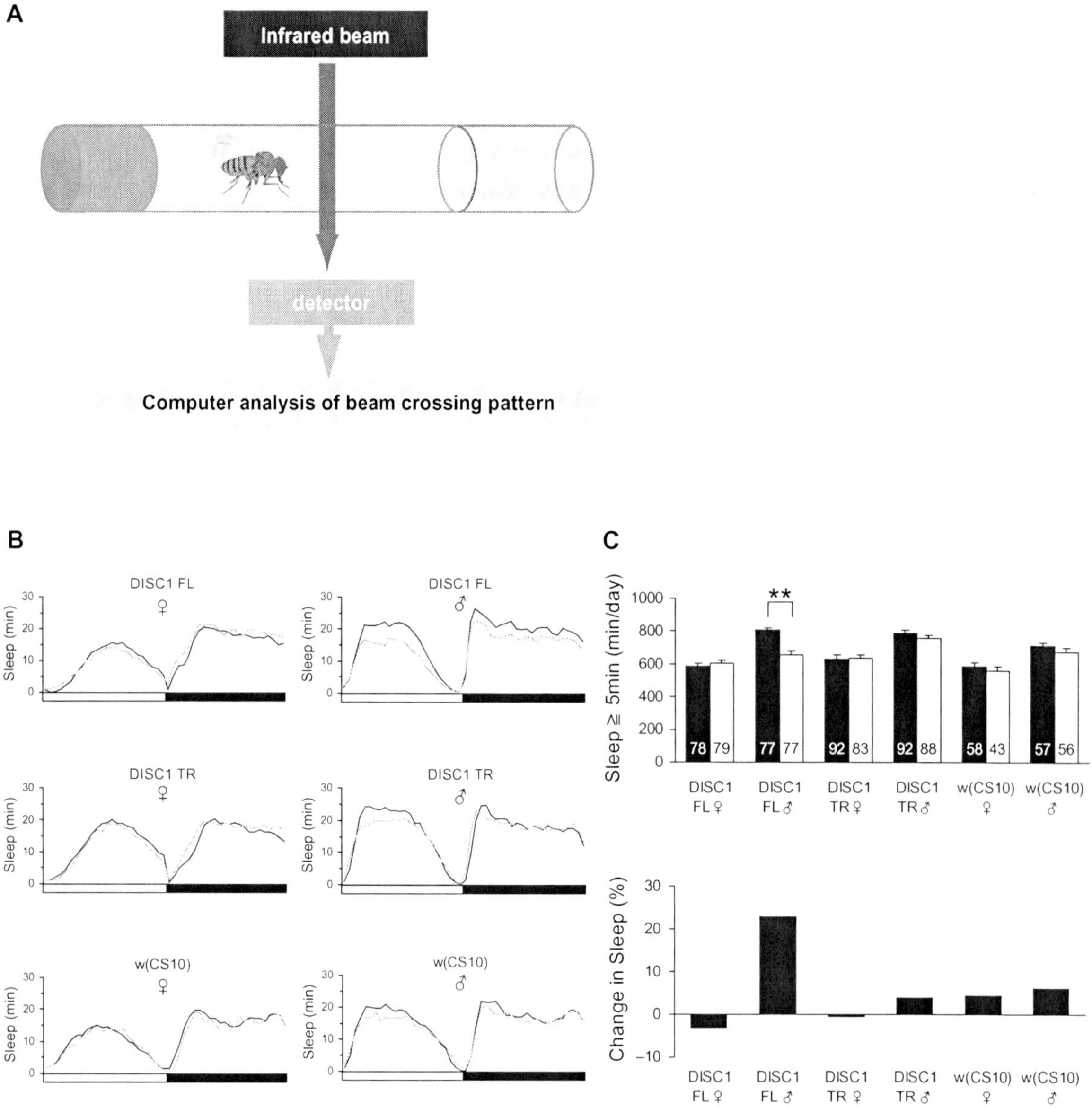

Fig. 3. DISC1 stimulates sleep amount in male flies. (A) Schematic illustration of *Drosophila* activity monitoring system. (B) Daily activity patterns of flies expressing DISC1-FL and DISC1(1-597) under control of the MB-GeneSwitch. Sleep patterns in flies with DISC1 expression (black lines) or without DISC1 expression (gray lines). Sleep was defined as a continuous rest state longer than 5 min. Average of three independent experiments. Note that the normal circadian activity pattern is not altered by DISC1 expression. (C) Total amount of sleep per day (24 h). Expression of DISC1-FL, but not DISC1(1-597) leads to significantly longer sleep in males. Black bar, +RU486. White bar, −RU486. Total number of flies used in each group is shown in the bar. The lower panel shows the % changes in sleep amount.

2000; Shaw et al., 2000; Andretic and Shaw, 2005; Ho and Sehgal, 2005; Zimmerman et al., 2008; Harbison et al., 2009). These behavioral paradigms provide unique opportunity to study the neuronal functions of schizophrenia susceptibility genes in this genetically highly tractable model system. In the past years, a large number of studies have highlighted functional convergence of multiple risk factors in the pathophysiological development of psychiatric disorders, and suggest that different risk factor genes confer psychiatric disease susceptibility by converging on shared pathophysiological processes that regulate the development and plasticity of the neural circuitries involved in higher-order cortical functions (Harrison and Weinberger, 2005; Jaaro-Peled et al., 2009). In addition to the studies of the cellular and molecular functions of individual susceptibility genes, *Drosophila* could be used as an attractive genetic system to study developmental and functional outcomes of convergence of multiple risk factors in living organisms. Mutant flies mutated for multiple risk factor genes are easily constructed, and multiple cognate human genes can be simultaneously expressed even in conjunction with mutant backgrounds for other risk factor genes. Finally, *Drosophila* allows identification of novel converging factors through systematic genetic screening for interacting genes that either suppress or enhance the observed abnormality caused by the cognate gene (Bier et al,. 2005).

Generating bona fide animal models for schizophrenia has a number of empirical and theoretical complications (Arguello and Gogos, 2006; Chen et al., 2006). When modeling a susceptibility gene in animals such as flies, the potential for genes to function differently in the unique contexts of the human and fly brains must be considered. Further, as is the case with other animal models, studies in flies cannot be complete without harnessing with knowledge obtained with vertebrate systems. However, recent advances in *Drosophila* neurogenetics and the accumulation of human psychiatric genetics data will aid the identification of novel risk factors, molecular and genetic components, and cellular processes that underlie each of the pathophysiology of psychiatric disorders, ultimately generating novel mechanism-based treatments.

Acknowledgments

Thanks to Drs Akira Sawa, Masami Shimoda, Norio Ishida, and Toshifumi Tomoda for their helpful discussion. This work was supported by Grants-in-Aid for Scientific Research from the Ministry of Education, Culture, Sport, Science, and Technology of Japan to K. F. T.

References

Adams, M. D., Celniker, S. E., Holt, R. A., Evans, C. A., Gocayne, J. D., Amanatides, P. G., et al. (2000). The genome sequence of *Drosophila melanogaster*. *Science*, *287*, 2185–2195.

Andretic, R., & Shaw, P. J. (2005). Essentials of sleep recordings in *Drosophila*: Moving beyond sleep time. *Methods in Enzymology*, *393*, 759–772.

Arguello, P. A., & Gogos, J. A. (2006). Modeling madness in mice: One piece at a time. *Neuron*, *52*, 179–196.

Auluck, P. K., Chan, H. Y., Trojanowski, J. Q., Lee, V. M., & Bonini, N. M. (2002). Chaperone suppression of alpha-synuclein toxicity in a *Drosophila* model for Parkinson's disease. *Science*, *295*, 865–868.

Berry, J., Krause, W. C., & Davis, R. L. (2008). Olfactory memory traces in *Drosophila*. *Progressin Brain Research*, *169*, 293–304.

Bier, E. (2005). Drosophila, the golden bug, emerges as a tool for human genetics. *Nature Reviews Genetics*, *6*, 9–23.

Bilen, J., & Bonini, N. M. (2005). Drosophila as a model for human neurodegenerative disease. *Annual Review of Genetics*, *39*, 153–171.

Bonini, N. M., & Fortini, M. E. (2003). Human neurodegenerative disease modeling using *Drosophila*. *Annual Reviewof Neuroscience*, *26*, 627–656.

Brand, A. H., & Perrimon, N. (1993). Targeted gene expression as a means of altering cell fates and generating dominant phenotypes. *Development*, *118*, 401–415.

Chen, J., Lipska, B. K., & Weinberger, D. R. (2006). Genetic mouse models of schizophrenia: From hypothesis-based to susceptibility gene-based models. *Biological Psychiatry*, *59*, 1180–1188.

Chubb, J. E., Bradshaw, N. J., Soares, D. C., Porteous, D. J., & Millar, J. K. (2008). The DISC locus in psychiatric illness. *Molecular Psychiatry*, *13*, 36–64.

Dietzl, G., Chen, D., Schnorrer, F., Su, K. C., Barinova, Y., Fellner, M., et al. (2007). A genome-wide transgenic RNAi library for conditional gene inactivation in *Drosophila*. *Nature*, *448*, 151–156.

Feany, M. B., & Bender, W. W. (2000). A *Drosophila* model of Parkinson's disease. *Nature*, *404*, 394–398.

Gottesman, II, & Gould, T. D (2003). The endophenotype concept in psychiatry: Etymology and strategic intentions. *Americal Journal of Psychiatry*, *160*, 636–645.

Greenspan, R. J., & van Swinderen, B. (2004). Cognitive consonance: Complex brain functions in the fruit fly and its relatives. *Trends in Neurosciences*, *27*, 707–711.

Hains, A. B., & Arnsten, A. F. (2008). Molecular mechanisms of stress-induced prefrontal cortical impairment: Implications for mental illness. *Learning & Memory*, *15*, 551–564.

Jaaro-Peled, H., Hayashi-Takagi, A., Seshadri, S., Kamiya, A., Brandon, N. J., Sawa, A. (2009) Neurodevelopmental mechanisms of schizophrenia: Understanding disturbed postnatal brain maturation through neuregulin-1–ErbB4 and DISC1. *Trends in Neurosciences*. *32*(9): 485–495. [Epub 2009 Aug 26]

Joiner, W. J., Crocker, A., White, B. H., & Sehgal, A. (2006). Sleep in *Drosophila* is regulated by adult mushroom bodies. *Nature*, *441*, 757–760.

Harbison, S. T., Mackay, T. F., & Anholt, R. R. (2009). Understanding the neurogenetics of sleep: Progress from *Drosophila*. *Trends in Genetics*, *25*, 262–269.

Harrison, P. J., & Weinberger, D. R. (2005). Schizophrenia genes, gene expression, and neuropathology: On the matter of their convergence. *Molecular Psychiatry*, *10*, 40–68.

Heisenberg, M. (2003). Mushroom body memoir: From maps to models. *Nature Reviews Neuroscience*, *4*, 266–275.

Hendricks, J. C., Finn, S. M., Panckeri, K. A., Chavkin, J., Williams, J. A., Sehgal, A., et al. (2000). Rest in *Drosophila* is a sleep-like state. *Neuron*, *25*, 129–138.

Hendricks, J. C., Williams, J. A., Panckeri, K., Kirk, D., Tello, M., Yin, J. C., et al. (2001). A non-circadian role for cAMP signaling and CREB activity in *Drosophila* rest homeostasis. *Nature Neuroscience*, *4*, 1108–1115.

Ho, K. S., & Sehgal, A. (2005). Drosophila melanogaster: An insect model for fundamental studies of sleep. *Methods in Enzymology*, *393*, 772–793.

Ishizuka, K., Paek, M., Kamiya, A., & Sawa, A. (2006). A review of Disrupted-In-Schizophrenia-1 (DISC1): Neurodevelopment, cognition, and mental conditions. *Biological Psychiatry*, *59*, 1189–1197.

Kamiya, A., Kubo, K., Tomoda, T., Takaki, M., Youn, R., Ozeki, Y., et al. (2005). A schizophrenia-associated mutation of DISC1 perturbs cerebral cortex development. *Nature Cell Biology*, *7*, 1167–1178.

Kilpinen, H., Ylisaukko-Oja, T., Hennah, W., Palo, O. M., Varilo, T., Vanhala, R., et al. (2008). Association of DISC1 with autism and Asperger syndrome. *Molecular Psychiatry*, *13*, 187–196.

Kirkpatrick, B., Xu, L., Cascella, N., Ozeki, Y., Sawa, A., & Roberts, R. C. (2006). DISC1 immunoreactivity at the light and ultrastructural level in the human neocortex. *Journalof Comparative Neurology*, *497*, 436–450.

Lessing, D., & Bonini, N. M. (2009). Maintaining the brain: Insight into human neurodegeneration from *Drosophila melanogaster* mutants. *Nature Reviews Genetics*,

Liu, L., Wolf, R., Ernst, R., & Heisenberg, M. (1999). Context generalization in *Drosophila* visual learning requires the mushroom bodies. *Nature*, *400*, 753–756.

Mao, Z., Roman, G., Zong, L., & Davis, R. L. (2004). Pharmacogenetic rescue in time and space of the rutabaga memory impairment by using Gene-Switch. *Proceedingsof the National Academyof Sciences of the United Statesof America*, *101*, 198–203.

Margulies, C., Tully, T., & Dubnau, J. (2005). Deconstructing memory in *Drosophila*. *Current Biology*, *15*, R700–713.

McGuire, S. E., Deshazer, M., & Davis, R. L. (2005). Thirty years of olfactory learning and memory research in *Drosophila melanogaster*. *Progressin Neurobiology*, *76*, 328–347.

McGuire, S. E., Roman, G., & Davis, R. L. (2004). Gene expression systems in *Drosophila*: A synthesis of time and space. *Trends in Genetics*, *20*, 384–391.

Millar, J. K., James, R., Christie, S., & Porteous, D. J. (2005). Disrupted in schizophrenia 1 (DISC1): Subcellular targeting and induction of ring mitochondria. *Molecular and Cellular Neurosciences*, *30*, 477–484.

Millar, J. K., Mackie, S., Clapcote, S. J., Murdoch, H., Pickard, B. S., Christie, S., et al. (2007). Disrupted in schizophrenia 1 and phosphodiesterase 4B: Towards an understanding of psychiatric illness. *Journal of Physiology*, *584*, 401–405.

Millar, J. K., Wilson-Annan, J. C., Anderson, S., Christie, S., Taylor, M. S., Semple, C. A., et al. (2000). Disruption of two novel genes by a translocation co-segregating with schizophrenia. *Human Molecular Genetics*, *9*, 1415–1423.

Morris, J. A., Kandpal, G., Ma, L., & Austin, C. P. (2003). DISC1 (Disrupted-In-Schizophrenia 1) is a centrosome-associated protein that interacts with MAP1A, MIPT3, ATF4/5 and NUDEL: Regulation and loss of interaction with mutation. *Human Molecular Genetics*, *12*, 1591–1608.

Neuser, K., Triphan, T., Mronz, M., Poeck, B., & Strauss, R. (2008). Analysis of a spatial orientation memory in *Drosophila*. *Nature*, *453*, 1244–1247.

O'Donovan, M. C., Craddock, N., Norton, N., Williams, H., Peirce, T., Moskvina, V., et al. (2008). Identification of loci associated with schizophrenia by genome-wide association and follow-up. *Nature Genetics*, *40*, 1053–1055.

Pitman, J. L., McGill, J. J., Keegan, K. P., & Allada, R. (2006). A dynamic role for the mushroom bodies in promoting sleep in *Drosophila*. *Nature*, *441*, 753–756.

Sawamura, N., Ando, T., Maruyama, Y., Fujimuro, M., Mochizuki, H., Honjo, K., et al. (2008). Nuclear DISC1 regulates CRE-mediated gene transcription and sleep homeostasis in the fruit fly. *Molecular Psychiatry*, *13*, 1138–1148, 1069.

Sawamura, N., Sawamura-Yamamoto, T., Ozeki, Y., Ross, C. A., & Sawa, A. (2005). A form of DISC1 enriched in nucleus: Altered subcellular distribution in orbitofrontal cortex in psychosis and substance/alcohol abuse. *Proceedingsof the National Academyof Sciences of the United Statesof America*, *102*, 1187–1192.

Shaw, P. J., Cirelli, C., Greenspan, R. J., & Tononi, G. (2000). Correlates of sleep and waking in *Drosophila melanogaster*. *Science*, *287*, 1834–1837.

St Clair, D., Blackwood, D., Muir, W., Carothers, A., Walker, M., Spowart, G., et al. (1990). Association within a family of a balanced autosomal translocation with major mental illness. *Lancet*, *336*, 13–16.

Tang, S., Wolf, R., Xu, S., & Heisenberg, M. (2004). Visual pattern recognition in *Drosophila* is invariant for retinal position. *Science*, *305*, 1020–1022.

Tweedie, S., Ashburner, M., Falls, K., Leyland, P., McQuilton, P., Marygold, S., et al. (2009). FlyBase: Enhancing Drosophila Gene Ontology annotations. *Nucleic Acids Research, 37*, D555–559.

van Swinderen, B. (2005). The remote roots of consciousness in fruit-fly selective attention? *Bioessays, 27*, 321–330.

van Swinderen, B. (2006). A succession of anesthetic endpoints in the *Drosophila* brain. *Journal of Neurobiology, 66*, 1195–1211.

van Swinderen, B., & Greenspan, R. J. (2003). Salience modulates 20–30 Hz brain activity in *Drosophila*. *Nature Neuroscience, 6*, 579–586.

van Swinderen, B., Nitz, D. A., & Greenspan, R. J. (2004). Uncoupling of brain activity from movement defines arousal States in *Drosophila*. *Current Biology, 14*, 81–87.

Venter, J. C., Adams, M. D., Myers, E. W., Li, P. W., Mural, R. J., Sutton, G. G., et al. (2001). The sequence of the human genome. *Science, 291*, 1304–1351.

Walsh, T., McClellan, J. M., McCarthy, S. E., Addington, A. M., Pierce, S. B., Cooper, G. M., et al. (2008). Rare structural variants disrupt multiple genes in neurodevelopmental pathways in schizophrenia. *Science, 320*, 539–543.

Wu, Y. R., Wang, C. K., Chen, C. M., Hsu, Y., Lin, S. J., Lin, Y. Y., et al. (2004). Analysis of heat-shock protein 70 gene polymorphisms and the risk of Parkinson's disease. *Human Genetics, 114*, 236–241.

Zimmerman, J. E., Naidoo, N., Raizen, D. M., & Pack, A. I. (2008). Conservation of sleep: Insights from non-mammalian model systems. *Trends in Neurosciences, 31*, 371–376.

A. Sawa (Ed.)
Progress in Brain Research, Vol. 179
ISSN 0079-6123

CHAPTER 13

Primate models of schizophrenia: future possibilities

Arthur A. Simen[a], Ralph DiLeone[a] and Amy F.T. Arnsten[b,*]

[a]*Department of Psychiatry, Yale University School of Medicine, New Haven, CT, USA*
[b]*Department of Neurobiology, Yale Medical School, New Haven, CT, USA*

Abstract: Schizophrenia is a disorder of the association cortices, with especially prominent structural and functional deficiencies in the dorsolateral prefrontal cortex (PFC). True dorsolateral PFC is found only in higher primates, and is characterized by highly elaborate pyramidal cells with extensive recurrent connections. The development of the primate PFC also involves distinct developmental and genetic pathways. Thus, primate models may be particularly important in determining the functional impact of genetic changes in patients with schizophrenia. Genes involved with pyramidal cell network connectivity may be especially important to study in primates, as their effects may be magnified in the extensively connected primate neurons. Adeno-associated virus technology appears particularly promising for studying the impact of genetic insults on the structure and function of the primate association cortex.

Keywords: Primate; prefrontal cortex; DISC1; evolution; working memory

Schizophrenia is a profound cognitive disorder with established neuropathological insults to the high-order association cortices that subserve thought. As the association cortices expand tremendously from rodents to monkeys to humans, nonhuman primate (NHP) models may be particularly important for understanding the impact of genetic insults on higher cortical circuitry, and their relevance to symptoms of schizophrenia. Recent advances in viral technologies may be especially useful in determining how genetic variants associated with schizophrenia alter higher-order cortical structure and function.

Schizophrenia as an affliction of the association cortices

Schizophrenia is characterized by fundamental cognitive deficits that significantly impede daily life and social relationships (Barch, 2005). Neuropathological and imaging studies have shown deficits in the structural integrity and functional activity of the frontal and temporal association cortices in patients with schizophrenia, particularly in the dorsolateral prefrontal cortex (PFC) (Lewis and Gonzalez-Burgos, 2006; Ragland et al., 2007). Neuropathological studies have revealed loss of dendritic spines (Selemon et al., 1995; Glantz and Lewis, 2000) from the layer III pyramidal cells in dorsolateral PFC that form the recurrent excitatory networks which subserve representational knowledge, our "mental sketchpad" (Goldman-Rakic, 1995).

*Corresponding author.
Tel.: +1 203 785 4431; Fax: +1 203 785 5263;
E-mail: amy.arnsten@yale.edu

DOI: 10.1016/S0079-6123(09)17913-X

Recent studies suggest that many of the so-called positive symptoms of schizophrenia such as hallucinations and delusions likely involve disruption of association cortical processing as well. For example, auditory hallucinations may arise from impaired communication between the PFC and temporal association cortices involved in language production and interpretation; specifically, weakened corollary discharge from the PFC to Wernicke's area to indicate that inner speech is self-generated (Ford et al., 2002). This idea has received experimental support from studies that used transmagnetic stimulation to weaken Wernicke's area that found a reduction in auditory hallucinations in patients with schizophrenia (Hoffman et al., 2005). Disruptions in PFC and temporal association cortices have also been associated with formal thought disorder, e.g. Perlstein et al. (2001). Most recently, the rostral dorsomedial PFC has been linked to reality testing and psychosis (Simons et al., 2008), while disturbances in the right dorsolateral PFC underlie delusional thinking (Corlett et al., 2007; Sun et al., 2008). Thus, disruptions in the higher association cortices are key to schizophrenia symptomatology. Altered cortical processing may be magnified by excessive dopamine (DA) D2 receptor signaling in the caudate, while medications with DA D2 blocking properties may diminish this amplification, but not solve the underlying cause of the illness. Thus, understanding genetic impacts on cortical circuitry is of highest priority in revealing the underlying etiology of schizophrenia.

The vast evolution of the association cortices from rodents to primates

The great, areal expansion of the association cortices

The association cortices expand greatly from rodents to monkeys to humans. As shown in Fig. 1, the cortex expands 100 times from mouse to monkey, and 1000 times from mouse to human, with the vast majority of this increase comprised of association cortex (Rakic, 2009). In contrast, the hippocampus has become relatively smaller in primates, as rodents and primates have taken differing and distant evolutionary paths (Fig. 1D). Indeed, there are types of neurons that are unique to primate and/or human cortex, and developmental trajectories unique to primate cortices (Rakic, 2009).

The development of the immensely convoluted human cortex begins in the second trimester of pregnancy, and is mediated by extraordinarily complex, yet precise, molecular events (for a recent review, see Rakic (2009)). Recent whole-genome, exon-level expression characterization of the developing human brain has revealed molecules that are differentially expressed in the developing primate and/or human association cortices (Johnson et al., 2009). The wealth of genes identified by this and similar studies will likely guide many future explorations of molecules mediating cortical development. Importantly, differential gene expression was prevalent even *within* the developing PFC.

As can be seen in Figs. 1 and 2, the PFC expands tremendously from rodents to primates, even on the medial surface (Fig. 2) where rodent PFC is concentrated. The region of PFC most afflicted in schizophrenia — the granular dorsolateral PFC- does not exist in rodents, or even in some lower primates (Preuss, 1995). Instead, the medial PFC in rodents shares greatest anatomical and functional homology with some medial PFC areas in monkey (Preuss, 1995).

The greatly increased complexity of primate PFC neurons

The pyramidal cells of the PFC also have an expanded dendritic architecture to support their numerous connections, and have about 70% more spines than the pyramidal cells of the primary sensory or motor cortices (Jacobs et al., 2001). Elston et al. (2006) have written that "the highly branched, spinous neurons in the human granular PFC may be a key component of human intelligence". Similarly, the pyramidal neurons in monkey dorsolateral PFC, which subserve recurrent network connections, have extensive dendritic arborization, with especially elaborate basal dendrites (e.g. Fig. 3). Careful measurements show that the basal dendrites of layer III pyramidal cells in monkey dorsolateral PFC are more than 3 times the length, and have more than 6

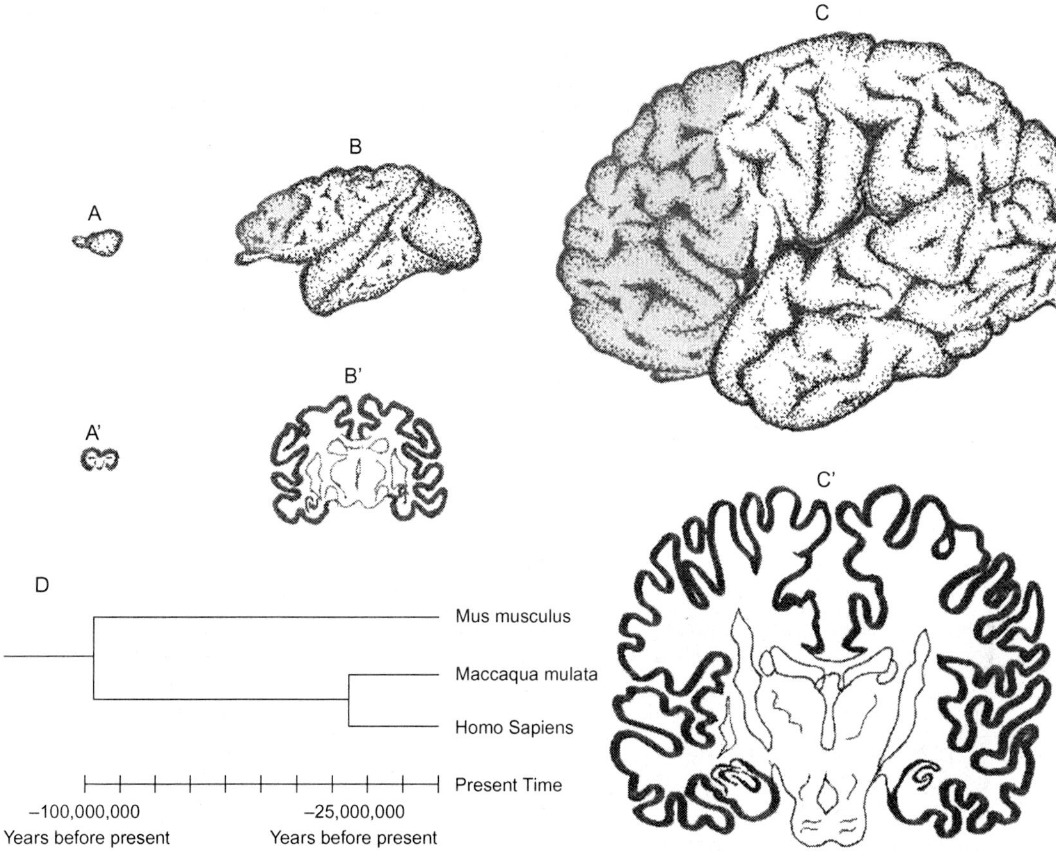

Fig. 1. Cerebral hemispheres of the mouse (A), macaque monkey (B), and human (C) drawn by Pasko Rakic at approximately the same scale to convey the overall difference in size and elaboration. The pink overlay indicates the area of the PFC that has no counterpart in mouse. The cross sections of the cerebral hemispheres of same species (A′, B′, C′) illustrate the relative small increase in the thickness of the neocortex compared to a large difference in surface of approximately 1:100:1000 X in mouse, macaque monkey, and human, respectively. (D) The timescale of phylogenetic divergence of mus musculus, maccaca mulata, and homo sapiens based on DNA sequencing data. Modified from Rakic (2009); figure courtesy of Dr Pasko Rakic, Department of Neurobiology, Yale Medical School.

times the complexity of basal dendrites of layer III pyramidal cells in rat medial PFC (Radley et al., 2005; Hao et al., 2007). Thus, developmental and/or genetic insults to synaptic connections may have particular impact on the function of dorsolateral PFC, and other high-order association cortex with elaborated dendritic complexity.

Pyramidal cell network connections as a key site of vulnerability in schizophrenia

Representational knowledge — our "mental sketchpad" — arises from recurrent excitation between PFC pyramidal neurons whose activity is sculpted by GABAergic interneurons (Goldman-Rakic, 1995). The interconnection of precise microcircuits is a fragile process, and many of the genetic alterations associated with schizophrenia appear to impact these network connections. These include genes key to the development of dendritic architecture (Lewis and Levitt, 2002), but also gene products needed for appropriate activity of the established circuit. For example, physiological studies in monkeys indicate that recurrent excitation between PFC pyramidal cells depends on *N*-methyl-D aspartate (NMDA) receptors (M. Wang and A. Arnsten, unpublished), and several

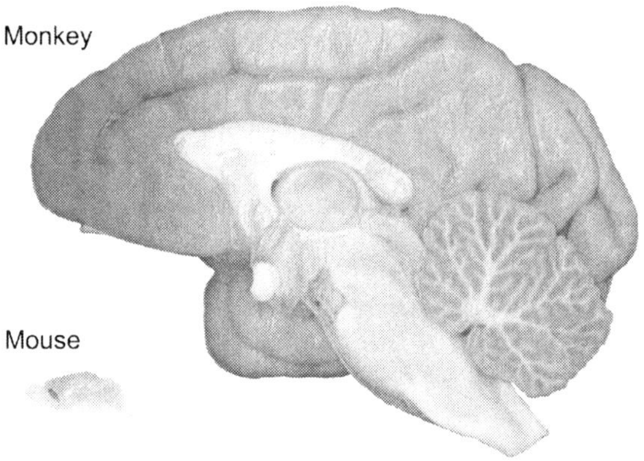

Fig. 2. A midsagittal view of the mouse and the monkey brain, with the medial PFC highlighted in red. The medial PFC in rodents is considered to be homologous to the medial PFC in primates, although the primate medial PFC is greatly expanded. Figure courtesy of Dr Nenad Sestan, Department of Neurobiology, Yale Medical School.

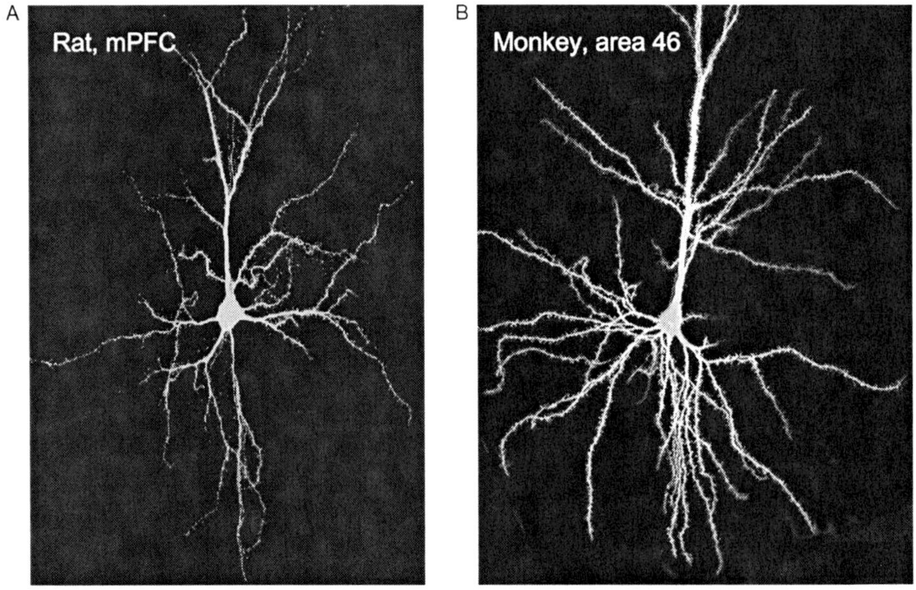

Fig. 3. The architecture of pyramidal cells greatly expands from rodents to primates. The basal dendrites of layer II/III pyramidal cells from rat prelimbic medial PFC (A) are much less extensive than the basal dendrites of dorsolateral PFC (area 46) pyramidal cells in monkeys (B). Neuron A is from Radley et al. (2005); Neuron B is from Hao et al. (2007). Figure courtesy of Dr John Morrison, Mt. Sinai School of Medicine.

genes associated with NMDA transmission have been linked to schizophrenia (Coyle et al., 2003). Disrupted In Schizophrenia (DISC1) and nicotinic α7 receptors have also been localized in PFC spines, and both have strong links to schizophrenia (Kirkpatrick et al., 2006; Duffy et al., 2009). DISC1 is thought to regulate cAMP levels, by activating PDE4 under conditions of high cAMP concentration (Millar et al., 2007). We have shown that excessive cAMP levels, e.g. during stress exposure, cause PFC network disconnection through opening of ion channels on dendritic spines (Wang et al.,

2007). In this context it is of interest that chronic stress exposure leads to spine loss in the PFC (Radley et al., 2005), especially in layer III, the layer harboring the recurrent microcircuits (Goldman-Rakic, 1995), that are most afflicted in schizophrenia (Glantz and Lewis, 2000). Future studies will uncover how genetic insults lead to the impaired development and/or degeneration of PFC microcircuitry.

The need for NHP models

Molecules involved in the development, maintenance, and physiology of dendrites and spines may have an especially large impact in primates compared to lower species, and have more effect on the association cortices than on sensorimotor functions. Thus, it will be important to study the impact of genetic alterations linked to mental illness in primate models. Indeed, evidence shows evolution of the molecules as well as the cortical circuitry, e.g. DISC1 is highly conserved between humans and monkeys but differs markedly from rodents (Bord et al., 2006). However, as transgenic approaches do not seem feasible in monkeys, alternative technologies will be needed. A promising direction for primate work is RNA interference through viral vectors.

Advent of viral technology

Highly efficient gene discovery analyses are revealing gene expression changes associated with PFC function and/or schizophrenia. The challenge that remains is how to most efficiently and effectively manipulate these genes to assess functional consequences in animal models. There are a number of approaches for manipulating gene expression including both transgenic and viral methods. Viral methods allow for efficient testing of gene function with spatial (e.g. specific brain region) and temporal (e.g. during development or in the adult) control. In the case of the PFC and schizophrenia, viral approaches offer the advantage of being effective across species, including NHPs that cannot be manipulated via transgenesis. Viral vectors have been long used to overexpress gene products in animal models, often with the goal of being applied as a gene therapy (Kaplitt et al., 1994). More recently, the vectors have been used to create genetic loss of function, or gene knockdown, models in animals. In this approach, the vectors are constructed to express short-hairpin RNA (shRNA) that targets specific genes, and then introduced to specific brain regions via stereotaxic surgery (Hommel et al., 2003). The shRNAs act to reduce mRNA and protein levels via RNA interference, or RNAi (Mello and Conte, 2004). RNAi allows for unprecedented control over specific gene expression. Importantly, RNAi can be used to target specific allelic variants or multiple genes at one time, making it amenable for mimicking human allelic states or evaluating effect of multiple mutations.

The utility of AAV vectors in modeling schizophrenia susceptibility

Understanding the pathophysiology of genetic polymorphisms that effect risk for psychiatric disease is challenging. Although simple model systems play an important role (Reinke and White, 2002; Davis, 2004), psychiatric disorders such as schizophrenia appear to involve complex changes in higher brain circuits (Lewis and Sweet, 2009), necessitating the study of higher mammalian species and NHPs in particular. Unfortunately, NHPs are not easily subject to genetic investigation due to a long generation time and difficulties with the manipulation of NHP embryos.

One technology that circumvents many of these difficulties is the use of recombinant viral vectors, adeno-associated virus (AAV) vectors (Aucoin et al., 2008; Buning et al., 2008). AAV is a single-stranded DNA parvovirus that efficiently targets dividing as well as nondividing cells. The genome consists of two inverted terminal repeats (ITRs) with a *rep* (replication related genes) and *cap* (capsid genes) open reading frame in between and it requires adenoviral genes for replication. The virus is limited to packaging genomes of about 4–5 kb in size. Because of the availability of systems with *rep*, *cap*, and adenoviral genes in *trans* and the fact that the virus is replication defective, the system is safe for laboratory personnel (ibid). AAV vectors have been successfully used in a variety of mammalian species from mice to humans, including NHPs, and can target a variety of cell types (see Table 1, Online Supplement).

A growing number of AAV serotypes have been isolated and characterized, and *cap* genes from one can be combined with rep genes from another to make "pseudotyped" viruses with altered properties (Grimm and Kay, 2003; Wu et al., 2006). Table 1 (online) summarizes the properties of various AAV viruses for work in CNS tissue. In addition, recombinant capsid genes have been generated (Girod et al., 1999; Shi et al., 2001; Maheshri et al., 2006; Koerber et al., 2007).

NHP species are a valuable resource and ensuring their safety is very important. AAV2 has been found to be an excellent vector for human gene therapy, and AAV9 also seems to be quite safe, but certain serotypes seem particularly prone to eliciting a strong host immune response, including AAV7 and AAV8 (Broekman et al., 2006; Klein et al., 2006; Howard et al., 2008). Immune reactivity toward the virus may limit the expression of the virus (Mingozzi and High, 2007; Zaiss and Muruve, 2008), and excess gene expression can lead to toxicity as described for AAV8 (Broekman et al., 2006; Klein et al., 2006). The recent development of neuronal cell lines from NHPs (Yasumoto et al., 2008) allows for in vitro testing prior to intracerebral injection.

AAV can mediate strong expression for very long periods of time although this is dependent on the promoter, AAV serotype, and host factors. AAV has been found to express up to 12 months or more after injection (Lo et al., 1999; Feng et al., 2004) based on rodent studies. Important advances in the efficiency of production and purification have been described (Aucoin et al., 2008) and large-scale production as is required for work in NHPs is therefore feasible.

Relevance to DISC1 function

DISC1 has emerged as one of the most important schizophrenia susceptibility genes. DISC1 was initially implicated in risk for schizophrenia when a Scottish family was identified with a translocation between chromosome 1 and chromosome 11 leading to a truncation of the DISC1 gene, reduced DISC1 expression, and strong linkage to schizophrenia and other serious mental illness (Wilson-Annan et al., 1997; Millar et al., 2001; Porteous and Millar, 2006). Single nucleotide polymorphisms in DISC1 were later found to be associated with schizophrenia but with much lower penetrance (Porteous et al., 2006). DISC1 interacts with a great many proteins including NDEL1 and PDE4B, and cAMP regulates some of these associations (Brandon, 2007). DISC1 shows very high conservation between human and rhesus monkey relative to rodent species (Bord et al., 2006), consistent with the need to study NHPs.

One important question with regard to DISC1 is whether the effects of polymorphisms depend on specific developmental periods. Research on children at high risk for schizophrenia has revealed that structural brain changes become evident during childhood and symptoms generally begin to develop in the later teen years (Sporn et al., 2003; Rapoport et al., 2005; Nugent et al., 2007; Arango et al., 2008). AAV expressing shRNAs against DISC1 could be used to model DISC1 loss of function at particular developmental time points to determine the developmental periods that depend most critically on DISC1 function.

DISC1 is ubiquitously expressed in brain tissue (Ma et al., 2002) (http://mouse.brain-map.org) as well as in certain peripheral tissues such as lymphocytes, macrophages, liver, and colonic epithelium (http://www.ncbi.nlm.nih.gov/sites/entrez?db=geo). Because AAV can be readily targeted to any brain region or peripheral tissue of interest, the technology allows the locus of DISC1 effects on behavioral function to be determined.

Recent work has determined that DISC1 undergoes extensive alternative splicing (Ma et al., 2002) (http://genome.ucsc.edu/). The consequence of this molecular diversity is currently not known. AAV technology can readily be used to overexpress particular DISC1 isoforms to allow their function to be determined. In addition, AAV technology could be used to express shRNA constructs that target specific DISC1 isoforms to determine which isoforms are most important for proper nervous system development and function.

Disadvantages vs. advantages of NHP model

Although NHP models will likely play a key role in elucidating the impact of genetic insults to higher cortical function, they have many serious disadvantages. NHP research is very expensive, tedious, and requires intensive regulation to maintain the highest ethical standards. It also proceeds very slowly, e.g. requiring 1–2 years to train monkeys to perform tasks required for physiological analyses. Thus, NHP research is not feasible for initial experimentation, and rodents will remain essential for the development of viral technologies or other methods that can be applied to NHP brain.

However, NHP models may be particularly important when weak or negative results are observed in rodent models. A genetic change may have a large impact on complex cortical circuitry in primates, but be scarcely evident in the simpler rodent cortex. Thus, we should always be cautious about negative results in rodent models, and encourage further research in more elaborate models when warranted. NHP models using irradiation (Selemon et al., 2005) or chronic amphetamine exposure (Selemon et al., 2007) have already provided insights regarding circuit changes, and AAV technology promises to further our understanding of genetic insults on PFC structure and function.

Summary

As schizophrenia is a disorder of the association cortices, primate models will be important for understanding the impact of genetic insults on the elaborate neuronal architecture and cortical circuitry distinct to the primate brain. The advent of AAV technology provides the opportunity to study genetic changes in primate brain, and will serve as an important bridge between rodent models and pathological profiles in patients, particularly when there have been negative findings in rodent models.

Conflicts of interest

Dr. Arnsten and Yale University have a license agreement with Shire Pharmaceuticals for the development of guanfacine for the treatment of ADHD. Dr Arnsten receives research funding, and consults, speaks, and teaches for Shire as well.

Acknowledgments

Some of the work discussed in this manuscript has been supported by a NARSAD Distinguished Investigator Award to A.F.T.A. A.S. was supported by NIA grant AG030970 and by the Claude D. Pepper Older Americans Independence Center at Yale University School of Medicine.

Appendix A. Supporting Information

Supplementary data associated with this article can be found in the online version at doi:10.1016/S0079-6123(09)17913-X.

References

Arango, C., Mareno, C., Martinez, S., Parellada, M., Desco, M., Moreno, D., et al. (2008). Longitudinal brain changes in early-onset psychosis. *Schizophr Bull, 34*, 341–353.

Aucoin, M. G., Perrier, M., & Kamen, A. A. (2008). Critical assessment of current adeno-associated viral vector production and quantification methods. *Biotechnology Advances, 26*, 73–88.

Barch, D. M. (2005). The cognitive neuroscience of schizophrenia. *Annual Review of Clinical Psychology, 1*, 321–353.

Bord, L., et al. (2006). Primate disrupted-in-schizophrenia-1 (DISC1): High divergence of a gene for major mental illnesses in recent evolutionary history. *Neuroscience Research, 56*, 286–293.

Brandon, N. J. (2007). Dissecting DISC1 function through protein–protein interactions. *Biochemical Society Transactions, 35*, 1283–1286.

Broekman, M. L., et al. (2006). Adeno-associated virus vectors serotyped with AAV8 capsid are more efficient than AAV-1 or −2 serotypes for widespread gene delivery to the neonatal mouse brain. *Neuroscience*, *138*, 501–510.
Buning, H., et al. (2008). Recent developments in adeno-associated virus vector technology. *Journal of Gene Medicine*, *10*, 717–733.
Corlett, P. R., et al. (2007). Disrupted prediction-error signal in psychosis: Evidence for an associative account of delusions. *Brain*, *130*(Pt 9), 2387–2400.
Coyle, J. T., et al. (2003). Converging evidence of NMDA receptor hypofunction in the pathophysiology of schizophrenia. *Annals of the New York Academy of Sciences*, *1003*, 318–327.
Davis, R. H. (2004). The age of model organisms. *Nature Reviews Genetics*, *5*, 69–76.
Duffy, A. M., et al. (2009). Spatial and intracellular relationships between the alpha7 nicotinic acetylcholine receptor and the vesicular acetylcholine transporter in the prefrontal cortex of rat and mouse. *Neuroscience*, *161*, 1091–1103.
Elston, G. N., et al. (2006). Specializations of the granular prefrontal cortex of primates: Implications for cognitive processing. *The Anatomical Record. Part A, Discoveries in Molecular, Cellular, and Evolutionary Biology*, *288*, 26–35.
Feng, X., et al. (2004). Adeno-associated viral vector-mediated ApoE expression in Alzheimer's disease mice: Low CNS immune response, long-term expression, and astrocyte specificity. *Frontiers in Bioscience*, *9*, 1540–1546.
Ford, J. M., et al. (2002). Reduced communication between frontal and temporal lobes during talking in schizophrenia. *Biological Psychiatry*, *51*, 485–492.
Girod, A., et al. (1999). Genetic capsid modifications allow efficient re-targeting of adeno-associated virus type 2. *Nature Medicine*, *5*, 1052–1056.
Glantz, L. A., & Lewis, D. A. (2000). Decreased dendritic spine density on prefrontal cortical pyramidal neurons in schizophrenia. *Archivesof General Psychiatry*, *57*, 65–73.
Goldman-Rakic, P. S. (1995). Cellular basis of working memory. *Neuron*, *14*, 477–485.
Grimm, D., & Kay, M. A. (2003). From virus evolution to vector revolution: use of naturally occurring serotypes of adeno-associated virus (AAV) as novel vectors for human gene therapy. *Curr Gene Ther*, *3*, 281–304.
Hao, J., et al. (2007). Interactive effects of age and estrogen on cognition and pyramidal neurons in monkey prefrontal cortex. *Proceedings of the National Academy of Sciences of the United States of America*, *104*, 11465–11470.
Hoffman, R. E., et al. (2005). Temporoparietal transcranial magnetic stimulation for auditory hallucinations: Safety, efficacy and moderators in a fifty patient sample. *Biological Psychiatry*, *58*, 97–104.
Hommel, J. D., et al. (2003). Local gene knockdown in the brain using viral-mediated RNA interference. *Nature Medicine*, *9*, 1539–1544.
Howard, D. B., et al. (2008). Tropism and toxicity of adeno-associated viral vector serotypes 1, (2, (5, (6, (7, (8, and 9 in rat neurons and glia in vitro. *Virology*, *372*, 24–34.
Jacobs, B., et al. (2001). Regional dendritic and spine variation in human cerebral cortex: A quantitative golgi study. *Cerebral Cortex*, *11*, 558–571.
Johnson, M. B., et al. (2009). Functional and evolutionary insights into human brain development through global transcriptome analysis. *Neuron*, *62*, 494–509.
Kaplitt, M. G., et al. (1994). Long-term gene expression and phenotypic correction using adeno-associated virus vectors in the mammalian brain. *Nature Genetics*, *8*, 148–154.
Kirkpatrick, B., et al. (2006). DISC1 immunoreactivity at the light and ultrastructural level in the human neocortex. *The Journal of Comparative Neurology*, *497*, 436–450.
Klein, R. L., et al. (2006). Efficient neuronal gene transfer with AAV8 leads to neurotoxic levels of tau or green fluorescent proteins. *Molecular Therapy*, *13*, 518–527.
Koerber, J. T., et al. (2007). Engineering adeno-associated virus for one-step purification via immobilized metal affinity chromatography. *Human Gene Therapy*, *18*, 367–378.
Lewis, D. A., & Gonzalez-Burgos, G. (2006). Pathophysiologically based treatment interventions in schizophrenia. *Nature Medicine*, *12*, 1016–1022.
Lewis, D. A., & Levitt, P. (2002). Schizophrenia as a disorder of neurodevelopment. *Annual Review of Neuroscience*, *25*, 409–432.
Lewis, D. A., & Sweet, R. A. (2009). Schizophrenia from a neural circuitry perspective: Advancing toward rational pharmacological therapies. *The Journal of Clinical Investigation*, *119*, 706–716.
Lo, W. D., et al. (1999). Adeno-associated virus-mediated gene transfer to the brain: Duration and modulation of expression. *Human Gene Therapy*, *10*, 201–213.
Ma, L., et al. (2002). Cloning and characterization of DISC1, the mouse ortholog of DISC1 (Disrupted-in-Schizophrenia 1). *Genomics*, *80*, 662–672.
Maheshri, N., et al. (2006). Directed evolution of adeno-associated virus yields enhanced gene delivery vectors. *Nature Biotechnology*, *24*, 198–204.
Mello, C. C., & Conte, D. J. (2004). Revealing the world of RNA interference. *Nature*, *431*, 338–342.
Millar, J. K., Christie, S., Anderson, S., Lawson, D., Hsiao Wei Loh, D., Devon, R. S., et al. (2001). Genomic structure and localisation within a linkage hotspot of Disrupted In Schizophrenia 1, a gene disrupted by a translocation segregating with schizophrenia. *Mol Psychiatry*, *6*, 173–178.
Millar, J. K., et al. (2007). Disrupted in schizophrenia 1 and phosphodiesterase 4B: Towards an understanding of psychiatric illness. *Journal of Physiology*, *584*, 401–405.
Mingozzi, F., & High, K. A. (2007). Immune responses to AAV in clinical trials. *Current Gene Therapy*, *7*, 316–324.
Nugent, T. F., 3rd, Herman, D. H., Ordonez, A., Greenstein, D., Hayashi, K. M., Lenane, M., et al. (2007). Dynamic mapping of hippocampal development in childhood onset schizophrenia. *Schizophr Res*, *90*, 62–70.
Perlstein, W. M., et al. (2001). Relation of prefrontal cortex dysfunction to working memory and symptoms in

schizophrenia. *American Journal of Psychiatry, 158*, 1105–1113.
Porteous, D. J., & Millar, J. K. (2006). Disrupted in schizophrenia 1: Building brains and memories. *Trends in Molecular Medicine, 12*, 255–261.
Porteous, D. J., et al. (2006). The genetics and biology of DISC1 – an emerging role in psychosis and cognition. *Biological Psychiatry, 60*, 123–131.
Preuss, T. (1995). Do rats have prefrontal cortex? The Rose–Woolsey–Akert program reconsidered. *Journal of Cognitive Neuroscience, 7*, 1–26.
Radley, J. J., et al. (2005). Reversibility of apical dendritic retraction in the rat medial prefrontal cortex following repeated stress. *Experimental Neurology, 196*, 199–203.
Ragland, J. D., et al. (2007). Neuroimaging of cognitive disability in schizophrenia: Search for a pathophysiological mechanism. *International Review of Psychiatry, 19*, 417–427.
Rakic, P. (2009). Evolution of the neocortex: a perspective from developmental biology. *Nat Rev Neurosci, 10*, 724–735.
Rapoport, J. L., et al. (2005). The neurodevelopmental model of schizophrenia: What can very early onset cases tell us? *Current Psychiatry Reports, 7*, 81–82.
Reinke, V., & White, K. P. (2002). Developmental genomic approaches in model organisms. *Annu Rev Genomics Hum Genet, 3*, 153–178.
Selemon, L. D., et al. (2007). Amphetamine sensitization alters dendritic morphology in prefrontal cortical pyramidal neurons in the non-human primate. *Neuropsychopharmacology, 32*, 919–931.
Selemon, L. D., et al. (1995). Abnormally high neuronal density in the schizophrenic cortex: A morphometric analysis of prefrontal area 9 and occipital area 17. *Archives of General Psychiatry, 52*, 805–818.
Selemon, L. D., et al. (2005). Direct and indirect effects of fetal irradiation on cortical gray and white matter volume in the macaque. *Biological Psychiatry, 57*, 83–90.
Shi, W., et al. (2001). Insertional mutagenesis of the adeno-associated virus type 2 (AAV2) capsid gene and generation of AAV2 vectors targeted to alternative cell-surface receptors. *Human Gene Therapy, 12*, 1697–1711.
Simons, J. S., et al. (2008). Separable forms of reality monitoring supported by the anterior prefrontal cortex. *Journal of Cognitive Neuroscience, 20*, 447–457.
Sun, D., et al. (2008). Progressive brain structural changes mapped as psychosis develops in "at risk" individuals. *Schizophrenia Research*,
Wang, M., et al. (2007). Alpha2A-adrenoceptor stimulation strengthens working memory networks by inhibiting cAMP-HCN channel signaling in prefrontal cortex. *Cell, 129*, 397–410.
Wilson-Annan, J. C., Blackwood, D. H., Muir, W., Millar, J. K., & Porteous, D. J. (1997). An allelic association study of two polymorphic markers in close proximity to a balanced translocation t(1:11) that co-segregates with mental illness. *Psychiatr Genet, 7*, 171–174.
Wu, Z., et al. (2006). Adeno-associated virus serotypes: Vector toolkit for human gene therapy. *Molecular Therapy, 14*, 316–327.
Yasumoto, F., et al. (2008). Genetic manipulation of brain cultures from the primates: A novel tool for molecular studies of neuropsychiatric disorders. *Molecular Psychiatry, 13*, 116–118.
Zaiss, A. K., & Muruve, D. A. (2008). Immunity to adeno-associated virus vectors in animals and humans: A continued challenge. *Gene Therapy, 15*, 808–816.

Subject Index